Biological Diversity and its Conservation

Biological Diversity and its Conservation

– Editor –

Dr. Dushyant Kumar Sharma
Head
Department of Zoology
Govt. P.G. College, Morena

&

Dr. R.P. Singh
Assistant Professor
Department of Botany
Govt. P.G. College, Morena

2018

Daya Publishing House®
A Division of
Astral International Pvt. Ltd.
New Delhi - 110 002

First Published, 2011
Reprinted, 2018

ISBN: 978-81-7035-738-4 (Hardbound)

Publisher's note:

Every possible effort has been made to ensure that the information contained in this book is accurate at the time of going to press, and the publisher and author cannot accept responsibility for any errors or omissions, however caused. No responsibility for loss or damage occasioned to any person acting, or refraining from action, as a result of the material in this publication can be accepted by the editor, the publisher or the author. The Publisher is not associated with any product or vendor mentioned in the book. The contents of this work are intended to further general scientific research, understanding and discussion only. Readers should consult with a specialist where appropriate.

Every effort has been made to trace the owners of copyright material used in this book, if any. The author and the publisher will be grateful for any omission brought to their notice for acknowledgement in the future editions of the book.

Published by : **Daya Publishing House**®
A Division of
Astral International Pvt. Ltd.
– ISO 9001:2008 Certified Company –
4736/23, Ansari Road, Darya Ganj
New Delhi-110 002
Ph. 011-43549197, 23278134
E-mail: info@astralint.com
Website: www.astralint.com

Dedicated to

OUR PARENTS

प्रोफेसर धीरेन्द्र पाल सिंह
कुलपति
Prof. D.P. Singh
Vice-Chancellor

काशी हिन्दू विश्वविद्यालय
वाराणसी - 221 005 (भारत)
BANARAS HINDU UNIVERSITY
(Established by Parliament by Notification No. 225 of 1916)
VARANASI - 221 005 (INDIA)

Foreword

Biological diversity refers to the variability among living organisms including diversity within species, between species and of ecosystems. This biological variation in nature, also known as biodiversity, is result of the evolutionary processes. Biodiversity plays a crucial role in providing a flow of goods and ecosystem services that contribute to human well-being.

Scientists are cataloguing and studying global biodiversity with the hope that they may better understand it. They now understand that human impacts on global biodiversity have been dramatic, resulting in unprecedented losses in global biodiversity at all levels, from genes and species to entire ecosystems. Biodiversity loss due to human activities is an issue of major concern as reductions in biodiversity can alter both the magnitude and the stability of ecosystem processes. Therefore, it is the duty of all of us to work together to conserve our natural heritage and biodiversity. It has to be understood that no biodiversity conservation programme can succeed without the awareness and involvement of general public.

Phones : (0542) 2368938 (O), 2368339 (R); Fax : (0542) 2369100 (O), 2369951 (R)
E-mail : vc_bhu@sify.com, vcbhu1 @gmail.com

Therefore, an integrated conservation and development programme is needed with a goal to enhance biodiversity conservation by addressing the needs, constraints and opportunities of local people.

I understand that the present book on *"Biological Diversity and Its Conservation"* will play an important role in this direction by providing background scientific information and case studies. The book contains both review articles and research papers which will be useful not only to the students and researchers but also to the policy planners and the public in general.

I hope that this book will contribute in broadening the knowledge base for the conservation of biological diversity in India.

(D.P. Singh)

Preface

Variation is the essence of life. It occurs everywhere and every moment in nature. In fact, variation is the beauty of nature. This variation and variability among organisms and ecosystems is biological diversity or biodiversity. It encompasses all species of plants, animals and microorganisms and ecosystems and ecological processes. We can not think a life without biodiversity. But due to the greed of mankind, there has been a threat to the biodiversity. Industrialization and urbanization have caused a great loss to biodiversity. Time has come when we have to think seriously about this. No method can be successful without the awareness of public and its involvement. The idea behind this book is to make the people aware about biological diversity.

The book is divided into two sections- *Section I* includes review articles by various academicians while *Section II* has research papers by scientists and academicians.

No good work is possible without the cooperation and involvement of a number of people. We are thankful to all the authors who have contributed in this book. It is a matter of great pleasure to publish their work in this book.

We are grateful to Professor D.P. Singh for writing *Forward* for our book. We express our sincere thanks to him. Thanks are due to Dr D.K. Sharma, Govt. P. G. College, Guna M.P. for his valuable suggestions.

We are thankful to all those who have directly or indirectly helped us in the completion of this book. We express our gratitude to our family members for their cooperation and inspiration.

We acknowledge our thanks to Daya Publishing House, New Delhi for publishing this book in this form.

We present this book with the hope that it will be able to achieve its mission–to make students and general people aware about biological diversity and they will also come forward and make their contribution in its conservation.

Dushyant Kumar Sharma

R.P. Singh

Contents

Section I–Research Papers

List of Contributors

Arnold, Rashmi
Department of Biotechnology & Microbiology, T.R.S. College, Rewa, Madhya Pradesh

Arya, Sandeep
Institute of Environment & Development Studies, Bundelkhand University, Jhansi, Uttar Pradesh

Awasthi, Aripta
Department of Biotechnology and Microbiology, T.R.S. College, Rewa, Madhya Pradesh

Bhadkariya, Rajeev Kumar
Govt. P.G. College, Morena, Madhya Pradesh

Choudhary, Anuradha
M.V.M. College, Bhopal, Madhya Pradesh

Choudhary, Vikram Singh
Govt. State Law College, Bhopal, Madhya Pradesh

Devi, Khomdram Nermeshori
Indira Gandhi Academy of Environmental Education Research and Ecoplanning (IGAEERE), Jiwaji University, Gwalior, Madhya Pradesh

Gupta, R.B.
Govt. S.M.S. Model Science College, Gwalior, Madhya Pradesh

Kanhere, R.R.
Govt. K.R.G. College, Gwalior, Madhya Pradesh

Kulshrestha, P.
Govt. K.R.G. College, Gwalior, Madhya Pradesh

Kumar, Ashok
St. John's College, Agra, Uttar Pradesh

Mahor, R.K.
Govt. K.R.G. College, Gwalior, Madhya Pradesh

Mehta, C.J.
Govt. K.R.J. College, Gwalior, Madhya Pradesh

Minakshi
Department of Environment Science and Engineering, Guru Jambheswar University of Science & Technology, Hissar, Haryana

Mishra, Deepak
Department of Biotechnology, Rajeev Gandhi College, Satana, Madhya Pradesh

Mishra, J.K.
Govt. P.G. College, Morena, Madhya Pradesh

Mishra, Manish
Indian Institute of Forest Management, Nehru Nagar, Bhopal, Madhya Pradesh

Pandey, Sadhna
Govt. K.R.G. College, Gwalior, Madhya Pradesh

Pathak, M.C.
Govt. P.G. College, Morena, Madhya Pradesh

Pathak, Sudhir
Govt. Girls College, Morena, Madhya Pradesh

Rajesh, Renu
Govt. Nehru Degree College, Ashoknagar, Madhya Pradesh

Sati, Vishwambhar Prasad
Govt. K.R.G. College, Gwalior, Madhya Pradesh

Sharma, Dushyant Kumar
Govt. P.G. College, Morena, Madhya Pradesh

Singh, B.R.
Govt. P.G. College, Morena, Madhya Pradesh

Singh, Keshav
Govt. Chhatrashal College, Pichhore, Shivpuri, Madhya Pradesh

Singh, Prashant Kumar
Department of Tourism Management, Indira Gandhi National Tribal University, Amarkantak, Madhya Pradesh

Singh, R.P.
Govt. P.G. College, Morena, Madhya Pradesh

Upadhyay, Mukulita
Govt. P.G. College, Morena, Madhya Pradesh

Section I
Review Articles

Chapter 1

Biological Diversity: Why to Conserve it?

☆ *Dushyant Kumar Sharma*

Biological diversity or biodiversity is the degree of variations of life forms. It includes variation among animals, plants, microorganisms, the genetic variation among them and all their complex assemblages of communities and ecosystems. Biodiversity is the totality of genes, species and ecosystems of a region. The total number of species on earth is estimated to be between 3 million to 100 million out of which only 1435662 species are identified all over the world. The term 'biodiversity' was first used by Raymond F. Dasmann in 1968. There are three levels of biodiversity-genetic, species and ecosystem diversity.

Genetic diversity refers to the variations of genes within the species. It includes all the different genes contained in all individual plants, animals, fungi and microorganisms. It occurs within a species as well as among species. *Species diversity* refers to the variety of species within a region. *Ecosystem diversity* refers to all the different habitats and biological communities within individual ecosystem.

Distribution of biodiversity is not uniform. It depends and varies according to the climate, altitude, soil structure and presence of other species in an area. Land biodiversity is greater than ocean biodiversity (Benton, 2001). A region with a high level of endemic species is called a biodiversity hotspot. India has two biodiversity hotspot- the Eastern Himalayas and Western Ghats.

Importance of Biodiversity

Though the term biodiversity may not be well understood to some of us but we all are benefited and dependent on the ecological services provided by biodiversity. Biological diversity is the very basis of human survival and economic development as it provides food, housing, clothing, medicine and industrial raw material and offers a potential for providing many more, yet unknown benefits to mankind. Biodiversity supports a variety of natural ecosystem processes and services. Ecological services like photosynthesis, air and water purification, recycling of nutrients, pollination and prevention of soil erosion are provided by biodiversity. Biodiversity plays very important role in human health. About 80 per cent of the world population depends on medicines obtained from nature (Behera *et al.*, 2008). A large number of drugs are obtained directly or indirectly from biological sources. A wide range of industrial materials are directly obtained from biological resources. Rubber, oil, fiber, building material, timber and paper all are obtained from biological resources. In fact, we cannot imagine a day, hour, or even a second when we are not dependent on biological diversity for our survival.

Biodiversity is also the source of non-material benefits like spiritual and aesthetic values, knowledge system, culture diversity and spiritual inspiration. It is a source of inspiration to musicians, painters, writers and other artists. Hobbies like gardening and animal keeping are possible because of biodiversity.

Loss of Biodiversity

Extinction of species and loss of biodiversity is the rule of nature which happens all the time. However, in recent years ever increasing loss of biodiversity has posed a serious threat to the survival of mankind. There have been five great episodes of extinction in the history of earth's life. The first was in Ordovician period (448 million years ago), second in Devonian (365 million years ago), third in

Permian (286 million years ago), fourth in Triassic (210 million years ago) and the fifth in the Cretaceous period (66 million years ago). These were believed to have been caused by climatic change (the first four) and by a giant meteorite crash (the fifth one). Today, human activities are the main cause, which are largely responsible for biodiversity loss. It is estimated that about 27000 species become extinct every year. If this goes on with the same rate, 30 per cent of world's species may be gone by the year 2050. The current extinction rate is 100 to 1000 times that of natural rate of extinction. The main causes of biodiversity loss are: habitat destruction, invasive species, pollution, population, overexploitation and climatic changes.

Destruction of the habitat of plants and animals is the most important cause of extinction of species. Physically larger species and those living at lower latitude or in the forests or oceans are more sensitive to reduction in habitat area (Drakare *et al.*, 2006). Habitat extinction compels the species to move where they find it difficult to adapt and this may ultimately lead to their extinction. Human activities like deforestation, pollution and over population are ultimately responsible for habitat destruction.

Introduction of exotic species is also responsible for the loss of biological diversity. The endemic and other local species may not be able to compete with the exotic species and are unable to survive. The exotic organisms may either be predator; parasite or they may simply outcompete indigenous species for food and shelter. Overexploitation, in the form of hunting of animals and plants for their commercial value is a major reason for reduction in biodiversity. Illegal wildlife trade is the single largest threat to biodiversity loss. Overpopulation of human and over consumption of natural resources are the root causes of all biodiversity loss.

Genetic pollution *i.e.* uncontrolled hybridization, introgression and genetic swamping can lead to threatening and replacement of endemic species. As soon as a better variety is developed, the same is distributed far and wide and brought in use. As a result the local/endemic species are discarded and their specific genes are lost.

Climatic change, again a consequence of human activities, has very adverse effects on biological diversity. Global warming affects plants, animals and microorganisms, both by changing their habitats and by direct effects of temperature. Rapid climatic changes could lead to higher number of diseases, landslide and forest fire, which may result in destruction of animals and plants. All organisms are adapted to a particular range of physical and chemical conditions. Climatic changes also affect species at cellular level. They can alter the genetic makeup of the cell and temperature can also increase the rate at which cell use energy. Thus it affects the physiology of the cell. Change in the climate has caused a danger to the survival of hundreds of plants and animals.

In addition to the above mentioned factors, inadequate knowledge, inefficient use of information and economic system and policies which fail to value the environment are also responsible for the loss of biological diversity.

Biodiversity of India

India is a mega diversity nation due to its rich floral and faunal wealth. The richness in biodiversity is due to immense variety of climatic and altitudinal conditions. It accounts for 7-8 per cent of the recorded species of the world species. In India about 130000 species of plants and animals have been documented so far. With over 45000 species of plants, India accounts for about 12 per cent of known world plant species. India is rich in faunal wealth and has nearly 85000 animal species which accounts for 7.25 per cent of world's total fauna. India is also known for its crop diversity, with about 320 closely related wild species of rice, pulses, millets, vegetables, fruits and fiber plants.

India has many endemic plant and animal species. Among plant species, endemism is estimated at 33 per cent with about 140 endemic species (Botanical survey of India 1983). 396 endemic higher vertebrate species have been identified. Endemism among mammals and birds is relatively low. Only about 44 mammalian species and 55 bird species are endemic in India. Endemism in the amphibian and reptilians fauna is high in India. India has 172 species of animals or 2.9 per cent of total number of threatened species (Groombridge, 1993).

Conservation of Biodiversity

Biodiversity conservation is of global concern. The United Nations declared the year 2010 as the International year of biodiversity. Both *in situ* and *ex situ* methods of biodiversity conservation are equally important. Preserving diversity through gene bank, seed bank and *in vitro* storage are effective methods for biodiversity conservation. In gene banks, the plant and animal materials are conserved and are available for breeding, reintroduction, research and other purposes. This method is useful for long living perennial trees and shrubs. Seed banks are the most efficient and effective method of *ex situ* conservation for sexually reproducing seeds under long term storage. There are a number of seed banks in the world with specialization in the nature of the collection, geographical area, taxonomic groups, wild plants, forestry trees etc. Removal of exotic species is another approach in biodiversity conservation. There should be restriction on introduction of exotic species without proper investigation.

Minimizing the use of chemical pesticides is another technique for the survival of biodiversity.

Preserving the habitat is the most important issue in the conservation of biodiversity. Destruction of habitat (deforestation) in the name of industrialization and urbanization, should be immediately checked and steps should be taken to restore the habitat of animals and plant species. Conservation of biodiversity through establishment of protected areas like National Park, Wild life sanctuary, Biosphere Reserves, Marine Reserves etc. are very effective in controlling the loss of biodiversity. Special care should be taken for the species which are threatened and at the verge of extinction. Efforts should be made to protect the indigenous genetic diversity.

The traditional beliefs and practices can, too, play an important role in biodiversity conservation. For example, *Ocimum santum* is commonly planted in houses in northern India and considered sacred herb and is also used for medicinal purpose. Gadgil and Berkes (1991) refer that various traditional ecosystem approaches require a belief system which includes a number of prescriptions for restrained resource use.

There is an urgent need to check the unsustainable exploitation of the biological diversity. This should be improved through appropriate legal and institutional system. Biodiversity conservation is a global issue and biodiversity loss affects each and every one. Thus, all countries should come forward and formulate proper programme to conserve biodiversity, develop resources for sustainability. A holistic approach is required for biodiversity conservation.

No conservation method is successful without the involvement of public. Public understanding of such issues is extremely important to integrate ecosystem conservation and rural development. People are the integral part of the ecosystem and thus, can play effective role in biodiversity conservation. Local people better understand the problems and they have better solutions. With constant association with the surrounding environment, they learn how to utilize many plant and animal species effectively for their day to day needs. Participatory Rural Appraisal technique may play an important role in planning the biodiversity conservation through people's participation.

Conclusion

Biological diversity is essential for the harmonious existence of life on earth. Any change in the system leads to major imbalance in the ecological cycle. The survival and well being of mankind depends on the well being of biodiversity. The conservation strategies must have a holistic approach which may lead to sustainable development. A collaborative approach is required by national agencies and regional bodies responsible for policy making, planning, research and development. New global agreements should be made and properly followed. If we want to live happily on this planet, we have to preserve and conserve biological diversity.

References

Behera, K.K., Sahoo, S. and Patra, S. (2008). Floristic and medicinal uses of some plants of chandaka denudated forest patches of Bhubaneswar, Orissa, India. *Ethnobot. Leaflets* 12: 1043-1053.

Benton, M. J. (2001). Biodiversity on land and in the sea *Geological Journal*, 36: 211–230

Botanical Survey of India (1983). *Flora and Vegetation of India–An Outline*. Botanical Survey of India, Howrah. 24 pp.

Dasmann, R. F. (1968). A Different Kind of Country. MacMillan Company, New York

Drakare S, Lennon J.L., Hillebrand H., (2006). *The imprint of the geographical, evolutionary and ecological context on species-area relationships* Ecology Letters 9 (2), 215–227.

Gadgil, M. and Berkes, F. (1991). Traditional resource management systems. Resource Management and Optimization 8 : 127-141

Government of India (1985). Research and Reference Division Ministry of Information and Broadcasting.

Groombridge, B. (1983). Comments on the rain forests of southwest India and their herpetofauna. Paper prepared for the Centenary Seminar of the Bombay Natural History Society, 6-10 December, 1983. 18 pp. Revised, January 1984.

Hunter, M. L. (1996). *Fundamentals of Conservation Biology*. Blackwell Science Inc., Cambridge, Massachusetts.

IUCN (1987). *Centres of Plant Diversity: A Guide and Strategy for their Conservation* (An outline of a book being prepared by the Joint IUCN-WWF Plants Conservation Programme and IUCN Threatened Plants Unit).

Khoshoo, T.N. (1993). Himalayan biodiversity conservation- An overview pp.5-35. In U.Dhar [ed]. Himalayan Biodiversity Conservation strategies. Gyanodaya Prakashan, Nainital.

Pillai, V.N.K. (1982). Status of wildlife conservation in states and union territories. In: Saharia, V.B. (Ed.), *Wildlife in India*. Natraj Publishers, Dehra Dun. Pp. 74-91.

Chapter 2

Loss of Biodiversity: An Overview

☆ *Renu Rajesh*

India is the seventh largest country in the world and Asia's second largest nation with an area of 3,287,263 square km. Physically the massive country is divided into following four relatively well defined regions–the Himalayan mountains, the Gangetic river plains, the southern (Deccan) plateau, and the islands of Lakshadweep, Andaman and Nicobar. The climate of India is dominated by the Asiatic monsoon, most importantly by rains from the south-west between June and October, and drier winds from the north between December and February. The climate is dry and hot from March to May.

A rich variety of wetland habitats is found in India. A total of 1,193 wetlands, covering an area of about 3,904,543 ha, were recorded in a preliminary inventory coordinated by the Department of Science and Technology, of which 572 were natural. India is distinct, not only because of its geography, history and culture but also because of the great diversity of its natural ecosystems. In India, forests ranges from evergreen tropical rain forests in the Andaman and Nicobar Islands, the Western Ghats, and the north-eastern states, to dry alpine scrub high in the Himalaya to the north. Between the two extremes, the country has semi-evergreen rain forests, deciduous monsoon forests, thorn forests, subtropical pine forests in the lower montane zone and temperate montane forests. A great wealth of biological diversity is found in forests, wetlands and in marine areas of India.

Endemic Species

India has many endemic plant and vertebrate species. Among plants, species endemism is estimated at 33 per cent with about 140 endemic genera but no endemic families. Endemism is rich in north-east India, the Western Ghats and the north-western and eastern Himalayas. A small pocket of local endemism also occurs in the Eastern Ghats. The Gangetic plains are generally poor in endemics. The Andaman and Nicobar Islands contribute at least 220 species to the endemic flora of India.

Five locations important for conservation action so far in India are: the Agastyamalai Hills, Silent Valley and New Amarambalam Reserve and Periyar National Park (all in the Western Ghats), and the Eastern and Western Himalaya. There are 396 known endemic higher vertebrate species. Only 44 species of Indian mammal have a range that is confined entirely to within Indian territorial limits. Four endemic species of conservation significance occur in the Western Ghats. They are the Lion-tailed macaque, Nilgiri leaf monkey (locally better known as Nilgiri langur), Brown palm civet and Nilgiri tahr. As far as birds are concerned, only 55 species are endemic to India. Their distribution is concentrated in areas of high rainfall, *viz.*, mainly in eastern India along the mountain chains, south-west India (the Western Ghats), and the Nicobar and Andaman Islands. Indian reptilian and amphibian fauna is high in endemism. There are around 187 endemic reptiles, and 110 endemic amphibian species. Eight amphibian genera are not found outside India.

Threatened Species

India contains 172 species of animal considered globally threatened by IUCN, or 2.9 per cent of the world's total number of threatened species. These include 53 species of mammals, 69 birds, 23 reptiles and 3 amphibians.

Biodiversity: An Introduction

Biodiversity reflects the number, variety and variability of living organisms. It includes diversity within species, between species, and among ecosystems. This diversity changes from one location to another and over time. Biodiversity is everywhere, both on land and in water. It includes all organisms, from microscopic bacteria to more complex plants and animals. Biodiversity plays an important role in the way ecosystems function and in the many services they provide. These services include nutrients and water cycling, soil formation and retention, resistance against invasive species, pollination of plants, regulation of climate, as well as pest and pollution control by ecosystems. Several aspects of human well-being including human health, social relations, and freedom of choice are negatively affected by loss of biodiversity.

Society tends to have various competing goals, many of which depend on biodiversity. When humans modify an ecosystem to improve a service it provides, this generally, also results in changes to other ecosystem services. For example, actions to increase food production can lead to reduced water availability for other uses. In the long term, the value of services lost may greatly exceed the short-term economic benefits that are gained from transforming ecosystems. Changes in ecosystems are harming many of the world's poorest people, who are the least able to adjust to these changes.

Biodiversity Loss

The current loss of biodiversity and the related changes in the environment are now faster than ever before and there is no sign of this process slowing down. Species extinction is a natural part of Earth's history but human activities have increased the extinction rate by at least 100 times compared to the natural rate. Biodiversity is declining rapidly due to natural or human-induced factors which tend to interact and amplify each other. Natural or human-induced factors that directly or indirectly cause a change in biodiversity are referred to as drivers. *Direct drivers that* explicitly influence ecosystem processes include land use change, climate change, invasive species, overexploitation, and pollution. *Indirect drivers*, such as changes in human population, incomes or lifestyle, operate more diffusely, by altering one or more direct drivers. It is easier to measure some direct drivers of change, for instance, fertilizer usage, water consumption, irrigation, and harvests. Measurement data are less readily available for non-native species, climate change, land cover conversion, and landscape fragmentation. Changes in biodiversity are result of combinations of drivers that work over time, on different scales, and that tend to amplify each other. For example, population and income growth combined with technological advances can lead to climate change.

Direct Drivers

Direct drivers causing biodiversity loss are habitat loss, habitat change, fragmentation of forests (small fragments of habitat can only support small populations that tend to be more vulnerable to extinction); invasive alien species that establish and spread outside their normal distribution; overexploitation of natural resources (overexploitation remains a serious threat to many species, such as marine fish and invertebrates, trees, and animals hunted for meat); pollution and excessive fertilizer use. *In terrestrial ecosystems*, the main driver has been land cover change such as the conversion of forest to agriculture. *In marine systems*, however, fishing, and particularly over fishing, has been the main drivers of biodiversity loss. In freshwater ecosystems, causes are water regime changes, such as those following the construction of large dams; invasive species, which can lead to species extinction; and pollution, such as high levels of nutrients.

Changes in climate show significant impacts on biodiversity and ecosystems. As climate change will become more severe, the harmful impacts on ecosystem are expected. Climate change is expected to exacerbate risks of extinctions, floods, droughts, population declines, and disease outbreaks. Recent changes in climate, such as warmer temperatures in certain regions, have affected species distributions, population sizes, and the timing of reproduction or migration events, as well as the frequency of pest and disease outbreaks. By the end of the century, climate change and its impacts may become the main direct driver of overall biodiversity loss.

Indirect Drivers

The main indirect drivers are changes in human population, change in economic activity, science and technology, as well as socio-political and cultural and religious factors.

The growth of human population and urbanization are imposing degrading pressures on biodiversity through following ways-

1. Encroachments into natural habitats.
2. Conversion of natural habitats into human settlements.
3. The increasing population is causing an increase in the demands of food which is met through the expansion of agriculture. This is done through clearing of natural forests, reclaiming of wetlands or through the fragmentation of vast areas of natural habitat.
4. Use of hybrid seeds and agro-chemicals like pesticides, fungicides, insecticides, rodenticides and hormones etc.

Poverty is another important factor which is responsible for the loss of biodiversity in the world because the poor are frequently forced to occupy the marginal lands and so to encroach upon the fragile ecosystems as is evident from the examples of Wetlands in Bangladesh; hill forests in India and Nepal; and Mangroves in Thailand that are ecologically disturbed due to their occupation by the poor people for settlements and agriculture. Thus both the poverty as well as affluence causes pressures on natural ecosystems that finally lead to greater degradation of resources, environment and ultimately the biodiversity. However, the growth of human population, either directly or indirectly, is one biggest cause of the losses experienced by the environment.

The increasing applications of *genetically engineered* microorganisms and their establishments in the natural habitats are causing potential risks to the existing plants and animals. Firstly, some traits of the genetically engineered microorganisms harm the species on which most of the natural organisms depend for their survival. Secondly, the mixing of the genetic stock and the subsequent loss caused by this event and the general competitive superiority of the genetically modified organisms lead to the degradation of biodiversity in a region.

Why Are Species Becoming Extinct?

Extinction means the irreversible loss of a species. Related to extinction is the ecological concept of extirpation, or the elimination of a given species from a geographic range. Prior to becoming extinct, a species is labeled endangered when their numbers become very low. One reason many animals are becoming, or have already become, extinct is human hunting and poaching. Another reason for species decline is habitat destruction via city development and logging. Habitat depletion is by far the greatest cause of extinction. Land is being taken from wildlife in order to erect more condominiums and shopping malls. Cities are constantly expanding to incorporate more people and businesses. Few efforts, if any, are made to relocate the animals or to reconstruct their habitat elsewhere. Logging is occurring throughout the world to supply home building, paper, fuel, and other products. Rivers are being damaged, as are the forests that are left barren and lifeless. Third main reason for wildlife extinction and endangerment is pollution from industries and vehicles. Landfills are becoming overfilled and many companies are using rivers as dumping grounds. Finally, the use of chemicals such as DDT and pesticides are to blame. Although DDT is no longer in use, its effects are still observed.

Further, varieties of Biological Resources are in International demands today. To meet this international demand illegal activities of smuggling of these resources are being done through illegal routes. Due to these malpractices some of the species of plants and animals have gone to the status of threatened, rare, endangered or critically endangered. These activities have rendered a large number of valuable species of plants and animals to the verge of extinction. The international trade in the body parts of animals is increasing rather more rapidly due to their high prices in the international markets. Where possible the trade of whole animals is also in the full swing *e.g.* Chiru is now at the verge of extinction due to its brutal killing for its fine hair (to prepare shahtoosh shawls). According to an estimate of Food and Agricultural Organization (FAO), from 4000 to 6000 species of medicinal plants are in the routes of international trade. Varieties of plants are in high international demand due to high value of their wood. The Asian Rattan Palm is in high international demand due to its wood for the making of furniture. The Wood's Cycad or the Encephallartos woodii has become extinct in the wild due to its severe exploitation for the medicinal purposes.

What's Wrong with Extinction?

Some of the important effects of extinction are: *First*, extinction of animal species leads to an imbalance in the food chain. It may cause population explosions of some species or the extinction of others. When this happens, animals may start competing for human resources or may wipe out the species, we rely on, for survival. *Second,* once an animal becomes extinct, it can never again be admired and appreciated by humans. Humans' very existence depends on wildlife survival. *Third,* destruction of plant life increases the amount of carbon dioxide in the air. As a result the greenhouse affect exacerbate. *Fourth,* when a species goes extinct, its genome, (the entire genetic information carried by that species in its DNA and hitherto capable of transmission to its descendants or of natural selection), is forever lost to the world. Thus it is not just the species that is lost, but its genome too. Hence, the possibility for further speciation is also lost.

Exposure to one threat often makes a species more susceptible to another. Multiple threats may have unexpectedly dramatic impacts on biodiversity. The extinction of species due to habitat loss can be rapid for some species, while it may take hundreds of years for others. Agricultural land expansion and forest cover shrinkage, particularly in developing countries, will lead to a continuing decline in local and global biodiversity, mainly as a result of habitat loss.

Human well-being will be affected by biodiversity loss, both, directly and indirectly. Direct effects include an increased risk of sudden environmental changes such as fisheries collapses, floods, droughts, wildfires, and disease. Changes will also affect human well-being indirectly, for instance in the form of conflicts due to scarcity of food and water resources.

Protected areas are an essential part of conservation programs. Sites for protected areas need to be carefully chosen, ensuring that all regional ecosystems are well represented, and the areas need to be well designed and effectively managed. Economic incentives can be provided to conserve biodiversity and to use ecosystem services sustainably. Prevention and early intervention are the most successful and cost-effective way of tackling invasive species. Once an invasive species has become established, its control and particularly its eradication are extremely difficult and costly.

International Programs and Conventions for Conservation

India participated in many international agreements and programs concerned with aspects of conservation of nature and sustainable development.

1. Legal instruments such as the Convention on Biological Diversity (CBD). India signed the Convention on Biological Diversity on 5th June 1992, rectified it on 18th February 1994 and brought it into force on 19th May 1994. In 2002, the Parties to the Convention on Biological Diversity agreed on a target to achieve significant reduction in the rate of biodiversity loss at the global, regional, and national level to the benefit of all life on earth by 2010.
2. Scientific programs such as the UNESCO's Man and the Biosphere Program, a global program of international scientific cooperation.
3. India became a party to Convention on International Trade in Endangered Species (CITES) on 18th October 1976. It provides data to the CITES secretariat on the trade of endangered species through its CITES Management Authority.
4. India ratified the World Heritage Convention in 1977 and since then five natural sites have been inscribed as areas of 'outstanding universal value'. These sites are:

   Kaziranga National Park, Keoladeo National Park, Manas National Park, Sundarbans National Park, Nanda Devi National Park.
5. India has been a contracted party to the Ramsar (Wetlands) Convention since 1st February 1982. India has now six sites covering some 192,973 hectares of important wetlands. These sites are; Chilka Lake, Keoladeo National Park, Wular Lake, Harike Lake, Loktak Lake, Sambhar Lake.

The agriculture, fishery, and forestry sectors are directly dependent on biodiversity and affect it directly. The private sector can make significant contributions, for example by adopting certain agricultural practices. Companies should show greater corporate responsibility and should prepare their own biodiversity action plans. Strong institutions at all levels are essential to support biodiversity conservation and the sustainable use of ecosystems. International agreements need to include enforcement measures. Most direct actions to halt or reduce biodiversity loss need to be taken at local or national level. Suitable laws and policies developed by central governments can enable local levels of government to provide incentives for sustainable resource management. Informing the society about the benefits of conserving biodiversity will help maximize the benefits to society.

However, current trends show no sign of a slowdown of biodiversity loss, and direct drivers of loss such as land use change and climate change are expected to increase further.

Since changes take place over different time frames, longer-term goals and targets are needed to guide policy and actions, along with short-term targets. Even on economic grounds alone, there is substantial scope for greater protection of biodiversity. Ultimately, however, the level of biodiversity that survives on Earth will be determined not just by considerations of usefulness but also by ethical concerns.

What Can We Do to Help?

There are many organizations that are fighting to preserve wildlife, including Greenpeace and the World Wildlife Fund etc. Many of these organizations are run by volunteers who help educate others, plant trees, grow them, rescue and rehabilitate injured and orphaned wildlife, and raise funds to help protect nature. Let us join one today, or start our own; write letters to government, companies, and newspapers voicing our concerns regarding the environment and our ideas for preserving it. Sometimes, only a few well-written, heart-felt words are needed to change the scenario. We can volunteer to educate people to develop writing and advertising material, replant trees, boycott animal tested products, help injured wildlife, boycott items made with animal skins or ivory, and reduce, reuse, recycle. We must do something. A true person of integrity and heart takes action for what he believes in.

References

Botanical Survey of India (1983). Flora and Vegetation of India–An Outline. Botanical Survey of India, Howrah.

FAO/UNEP (1981). Tropical forest resources assessment project. Technical report No. 3.FAO, Rome.

Groombridge B (1993). The 1994 IUCN Red List of Threatened Animals. IUCN, Gland, Switzerland and Cambridge, UK.

ICBP (1992). Putting biodiversity on the map: priority areas for global conservation. International Council for Bird Preservation, Cambridge, UK.

Lal JB (1989). India's Forests: Myth and Reality. Natraj Publishers, New Delhi, India.

Nayar MP, Sastry ARK (1987). Red Data Book of Indian Plants, Vol. 1. Botanical Survey of India, Calcutta. MacKinnon, J. and MacKinnon, K. (1986). Review of the Protected Areas System in the Indo-Malayan Realm. International Union for the Conservation of Nature and Natural Resources, Gland, Switzerland and Cambridge, U.K.

Millenium Ecosystem Assessment, Ecosystem and Human Well-being: Biodiversity Synthesis (2005).

Scott, D.A. (1989). A Directory of Asian Wetlands. IUCN, Gland, Switzerland, and Cambridge, UK.

Chapter 3

Eco-Tourism: Linking Tourism and its Impact on Biological Diversity

☆ *Prashant Kumar Singh*

The concept of tourism as phenomenon involves the movement of people with in their own country or across the national borders. Tourism is concerned with pleasure, leisure, holidays and travel. These are the motivations that make people leave their normal place of work and residence for short-term temporary visits to other places. Tourism and travel are not synonyms. All tourism involves travel but all travel is not tourism. All tourism involves recreation but all recreation is not tourism. All tourism occurs during leisure time but all leisure time is not given to tourist pursuits.

As per World Tourism Organization (WTO), Tourism comprises the activities of persons traveling to and staying in places outside their usual environment for not more than one consecutive year for leisure, business and other purposes.

- ☆ The term "usual environment" is intended to exclude trips within the place of residence, trip to usual place of work or education and daily shopping and other local day-to-day activities.
- ☆ The threshold of twelve months is intended to exclude long-term migration.
- ☆ For the distance traveled there is no consensus. It varies from at least 40 km to at least 160 km away from home one way for any purposes other than commuting to work.

On analysis of above definition, we find the following features of tourism:

1. Tourism arises from the movement of people to, and their stay in various destinations.
2. There are two elements in all tourism – the journey to the destination and the stay.
3. The journey and the stay should take place outside the normal place of residence and work, so that tourism gives rise to activities.
4. The movement to destination is of temporary character with the intention of returning with in few days, weeks and months.
5. Destinations are visited for purposes other than taking up permanent residence or employment.

Tourism is a leisure activity which involves time and money and recreation is the main purpose for tourism. Attraction, accessibility and amenities at destination are three major basic components of tourism. Attraction is the principal resource of any destination. These resources are broadly classified as:

- ☆ Natural Resources
- ☆ Cultural Resources

Natural resources are key elements in a destination's attraction. The important natural resources are:

- ☆ Countryside
- ☆ Climate – Temperature, Rains, Snowfall, Sunshine.
- ☆ Natural Beauty – Landforms, Hills, Rocks, Gorges, Terrain.

- ✰ Water – Lakes, Ponds, Rivers, Waterfalls, Springs.
- ✰ Flora and Fauna – Biosphere Reserves.
- ✰ Wildlife – National Parks, Wildlife Sanctuaries.
- ✰ Beaches.
- ✰ Islands.
- ✰ Scenic Attractions.

These natural resources constitute a form of tourism which is called Nature Tourism. Wildlife Tourism, Mountain Tourism, Beach Tourism, Island Tourism, Rural Tourism, etc are various aspects of Nature Tourism. The diversity of wildlife in India is as rich its flora and fauna. There are about 88 National Parks, 490 Wildlife Sanctuaries and 12 Biosphere Reserves.

Indian Biodiversity Scenario

Natural resources, ecology, environment and biological diversity are closely connected with each other. A concise definition of biodiversity is "the totality of genes, species, and ecosystem in a region."

According to U.S. Office of Technology Assessment (1987), biological diversity is "the variety and variability among living organisms and ecological complexes in which they occur."

Biodiversity is the essential characteristic of planet earth. The Indian subcontinent is blessed with rich biological diversity. India is one of the twelve mega- biodiversity countries of the world. India occupies a unique position among global biodiversity. A large number of species are native to India. It is stated among the top ten nations of the world for its great diversity of plant life, especially flowering plants. About five thousand species of flowering plants belonging to one hundred forty one genera and forty seven families had birth in India. We are equally rich in insect, amphibian, reptiles, bird and mammalian species. Many of them are endemic to India, found nowhere else in the world.

Endemic species of both plants and animals are mostly found in North-East, Western Ghats and Andaman and Nicobar Island. In Western Ghats and North-East Himalayas about 1500 and 2000 species of plants and animals respectively are endemic.

Major Impacts of Tourisms on Natural Environment and Biodiversity

Although Tourism is Smokeless Industry but it has some environmental implications. As soon as, tourism operations take place the environment is inevitably changed. There is a complex interaction between tourism and environment. This is true that the tourism is damaging the environment. Tourism operations have some direct and indirect impacts on environment, biodiversity, ecology. The impacts can be positive or negative.

The direct environmental impacts of tourism include negative effects showing in Figure 3.1.

According to European Environment Agency tourism creates following environmental problems:

- ✰ Waste disposal
- ✰ Reducing level of biodiversity
- ✰ Pollution of inland waters
- ✰ Pollution of marine and costal zones.

The direct environmental impacts of tourism include following positive effects.

- ✰ Creation of National Parks and Wildlife Sanctuaries.
- ✰ Maintenance of forest.
- ✰ Preservation and restoration of historical buildings.

Eco-Tourism

The environmental impact associated with tourism can also be considered in terms of their direct, indirect and induced effects. It is not possible to develop tourism without incurring impacts on environment, ecology and biological diversity, but it is possible, with correct planning, to manage tourism development in order to minimize the negative impacts while encouraging positive impacts. This can be possible by Eco-tourism.

Eco-Tourism has been defined by World Tourism Organization (WTO) as, " tourism that involves traveling to relatively undisturbed natural areas with the specified object of studying, admiring and enjoying nature and its wild plants and animals, as well as existing cultural aspects (both of the past and present) found in these areas."

Figure 3.1: Effect of Tourism on Environment and Biodiversity

ATMOSPHERE

⇩

TOURISM OPERATIONS		IMPACT
Increase in travel to destination (Motor-car, Ship, Train, Aeroplane)	⇨	Air and Noise Pollution, Adverse impact on plant life, Loss of Recreation Value. Disturbance to nesting birds.

WATER

⇩

TOURISM OPERATIONS		IMPACT
Release garbage, sewage into lakes, river, beaches	⇨	Contamination, health hazards, destruction of aquatic plant and animal life, deterioration of ecological diversity
Release of oils from cruise ships, ferry boats, etc.		Increasing toxicity in water bodies, contaminated sea food, etc.

VEGETATION

⇩

TOURISM OPERATIONS		IMPACT
Chopping of trees for resort construction, Thoughtless use of fire in parks and forest, Vehicular traffic in parks and forests, collection of flower plants and fungi	⇨	Loss of forest wealth, Habitat loss and fragmentation erosion Constant trampling of plants, Conflagrations and forest areas. Adverse effect of plant life

In simple words, eco-tourism means an overall management of environment with a balance between requirements of tourism, ecology along with generating revenue for local people taking care not to alter the integrity of the ecosystem. The following are main characteristics of eco-tourism:

- ☆ Minimum environmental impact.
- ☆ Maximum economic benefit to local communities.
- ☆ Awareness about environment among local community.
- ☆ Maximum recreational satisfaction to participating tourists.

Forest act themselves as ecosystems and prevent soil erosion, water recharging, conservation of biodiversity, stabilizes sand dunes and prevent land slides in mountainous areas.

Finally we can say tourism development can become a positive factor for improving environment, ecology and biological diversity if good environmental planning is done. The quality of tourism product depends upon a high quality natural environment. Then through eco-tourism concept and good environmental planning, the development of tourism will not degrade the environment, ecology and biological diversity, in fact it can be improved.

References

Bhardwaj D.S., Kamra K.K, Choudhary Manjula, Kumar Ravi Bhusan, Boora S.S., Chand Mohinder, Taxak R.H. (2006). International Tourism Issues and Challenges, Kanishka Publishers, New Delhi.

Bhatia AK (2006). The business of Tourism Concept and Strategies, Sterling Publication, New Delhi.

Gupta S.P., Lal Krishna, Bhattacharyya Mahua (2002). Cultural Tourism in India, D.K. Print world, New Delhi.

Kumar Kapil and Course Team, History faculty IGNOU (2008). Tourism Impact, Foundation Course in Tourism, Viba Press Pvt. Ltd.

Sharma PD (2010). Ecology and Environment, Rastogi Publication, Meerut.

Sinha BK, Choudhary Shriti (2008). Environment, Pollution and Health Hazards, APH Publishing Corporation, New Delhi.

Chapter 4

Role of Women in Conservation and Management of Biodiversity

☆ *R.B. Gupta*

Man and women have distinct realms of knowledge and application for natural resource management both of which are necessary for sustainable use and conservation. Very little progress has been made in understanding the fundamental roles that women play in managing and conserving biodiversity. It is essential to recognize that woman has particular needs, interests and aspirations and that she make different contributions to the conservation and sustainable management of biodiversity.

Biodiversity of species and breeds has been and continues to be increased and cultivated through the work of women. Across the globe, women are home gardeners, pleasant farmers, plant domesticators, collectors of forest fruits, wild herbs and forest feed. They are live stock owners, dairy farmers, fisher-women and often also grocers.

There is a need for the full participation of women at all levels of policy making and implementation for the biological diversity conservation.

Biodiversity is a term we use to describe the variety of life on earth. It refers to the wide variety of ecosystem and living organism: animals, plants, their habitats and their genes. Biodiversity is the foundation of life on earth. It is crucial for the functioning of ecosystem which provides us with products and services without which we could not live. Oxygen, food, fresh water, fertile soil, medicine, shelter, protection from storms and floods, stable climate and recreation-all have their source in nature and healthy ecosystem. But biodiversity gives us much more than this. We depend on it for our security and health, it strongly affects our social relations and gives us freedom and choice. Biodiversity is extremely complex, dynamic and varied like no other feature of the earth.

Status of Women in Society

Women are part of the society. They contribute in every aspect of life and their contributions are validated. To provide role models, young women do need to have role models. Women have been oppressed for thousands of years reaching into present day. Women, today certainly enjoy more political and social freedom than women of the past.

Western cultures were dominated by men but the influence of women was allowed to have a greater impact, bringing about women's right at a greater pace. Besides, it brings an identity of the culture itself. This may have come about because of the role women played in the industrial world. After world war one and two, in many countries women demanded equal pay scales, greater employment and equal status in society.

Role in Biodiversity Protection

The role of women are distinctly different from man. Naturally their perspective and understanding of biodiversity are different. Women have a natural stake in protecting biodiversity, it has not been recognized. Their role in conservation has been overlooked. The loss of habitats and biodiversity mostly affects the under privileged and many of them are women.

There is an urgent need for greater diversity within global organizations. Whether it is developed or emerging economies, the role of women is critical to economic, social and cultural growth and there is no escaping the fact that greater diversity today is a business need. Nations need to focus all the mere on inclusively and ensuring that a large number of women become a part of the work force. Workforce dynamics are undergoing a change with western nations facing issues of aging workforce, significant shortfalls of talent and declining population rates adding to the concern of addressing global war for talent.

Research by the National Sample Survey Organization (NSSO) on India reveals that a significant increase in women on the workforce can result in 12 per cent increase in per capita income. In Chile and around the world, indigenous rural women have been experts in sustainable development since long before the term existed. Dr. Gloria Montenegro shows how combining 21st century science with age old folk knowledge can help save the environment, create new medicines, and alleviate poverty in the under developed regions.

For many landless women, the only source of income is often just the forests produce. Women are dependent on forests for food, fuel, fodder, medicines, fibers, and so on. Moving in and out of forest, women have down the years accumulated loads of knowledge in the sustainable use of natural resources as well as numerous insights of the value of biodiversity.

Rural women who spin sheep's wool (work done mostly by women) provide an interesting example. These women know the properties of different plant species they use to dye their wool fibers and they know which parts of plant must be used to obtain the desired color. They are careful to take only the parts of the plant that they need and they harvest it in a manner that ensures re-growth. Similar attitude is found in women beekeepers.They care for and protect various different species, but they pay particular attention to the species from which honey bees collect nectar and pollens.

P V Sathesh, director Deccan Development society Hyderabad says "Women are the final caretakers of genetic and species diversity in agriculture. The knowledge about the seeds and the way it should be stemmed and sown is in the hand of women and not in man". The knowledge of the necessary growing conditions and nutritional character of various species gives women a unique crucial depository of experiences in seed collection and plant breeding. It is this knowledge that helps women to maintain the genetic diversity required to adapt to fluctuating weather patterns to ensure the survival of traditional crops.

Dr. S K Pandey (former director general of forest) pointed out that a rural India has a women centric agriculture system. Nothing can happen without them. Women are a great torch-bearer of biodiversity knowledge as they intimately understand it. If animals are sick they know what medicinal plants have to be administered? Women have the greatest concern as far as protecting biodiversity is concerned and they have to be brought in if we are serious about conservation.

Today, over 10 per cent of India's recorded wild flora and fauna are on the threatened list. Many are on the verge of extinction. In the last few decade, India has lost over 50 per cent of its forests and polluted over 70 per cent of its water bodies. It has built or cultivated over most of its grasslands. It has degraded most of its prescient coasts. As if this was not enough, animals like the tiger and rhino were hunted down. The green revolution opened the door for chemical fertilizers and pesticides and due to their overdose poisoning of land started. India is today not only the largest producer of pesticides; it is also the largest consumer. All this has taken heavy toll on biodiversity.

The role of women in traditional management practices has increasingly been appreciated globally as a strong incentive for biodiversity conservation. That role has high potential in enhancing conservation and sustainable use of natural resources including home gardens and therefore as a remedy for numerous forest conservation problems.

Across the globe, women predominate as wild plant gather, home gardeners and plant domestications, herbalist and seed custodians. Research on 60 home gardeners in Thailand revealed 230 different species, many of which had been rescued by women from neighboring forest before being cleared. Women in different regions of Latin America, Asia and Africa manage the interface between wild and domesticated species of edible plants. This role dates back to 1500–1900 B.C.

Women and men have different knowledge about the preference for plants and animals. For example, women criteria for choosing certain food crops may include cooking time, meal quality, taste, resistance to bind damage and ease of collection, processing, preservation and storage. Men are more likely to consider yield, suitability for a range of soil types and ease of storage. Both are essential for human welfare. It is evident that women often have specialized knowledge about 'neglected' species. The various studies demonstrate following conclusions:

- Women are mostly involved in home garden management related activities.
- Women are interested in conserving home gardens because they obtain such substantial benefits such as food security, income, healthcare and environmental benefits.
- Women were found to be aware of home garden conservation and tuned to motivating husbands, children and neighbors to conserve the agro biodiversity of home gardens. Findings suggest that increased involvement of women in a broad range of home garden management activities is not only beneficial for their own socio-economic well being but also imperative for sustaining the livelihood of their communities and for preserving the agro biodiversity in home gardens.

Conclusion

The present study has portrayed the role of women in the use, conservation and traditional management practices in the home gardens, which can be used as an entry point to build an economically viable and ecologically sustainable home garden management system. Based on the present study, the following general recommendations are suggested:

1. The role of women in every sphere of conservation should be measured and necessary support should be provided to ensure sustainable home garden conservation.
2. Clear government policies, national guidelines, strategies and plans for the involvement of women should be formulated and implemented.
3. The dissemination of technical information should target

women as they are the drivers of home garden management.

4. Special attention should be given to the significance of women's ancestral knowledge of biodiversity.
5. Recognizing and valuing women's knowledge and practices related to biodiversity helps us to understand the crucial step of effective participation for women in decision making for biodiversity conservation.

References

Dolan CS The "good wife" struggle over resources in the Kenyan horticulture sector. *The J. of Development Studies*, London, England.

Eyzaquirre P (2001). Growing diversity a Handbook for applying Ethnobetery to conservation and community development, September Vol. 7, IPG RI, Rome.

Pillai K Rajasekhara, B Suchintha (2006). Men empowerment for biodiversity conservation. *Indian J. of Agricultural Resources*, Vol. 5(4):338-355.

Paola Deda and Reneta Rerbian (2004). Women and biodiversity. The long journey from users to policy makers. *Natural Resources Forum*, 28:201-204.

Sayma Akhtar (2010). Role of women in traditional farming system as practiced in home gardens. *Tropical Conservation Science*, Vol. 3 (1):17-30.

Soar J (2002). Women, men and environmental change: gender dimensions of environmental policies and programmes. Population references bureau services.

Vanaja Ramprasad (1999). Women and biodiversity conservation. COMPASS news letter.

Chapter 5

Wetland Conservation: A Review

☆ *Sandeep Arya & Minakshi*

Wetlands are areas where water is the primary factor controlling the environment and the associated plant and animal life. They occur where the water table is at or near the surface of the land, or where the land is covered by water. Wetlands, natural and manmade, freshwater or brackish, provide numerous ecological services. Wetlands are often biodiversity 'hotspots' (Reid *et al.*, 2005), as well as functioning as filters for pollutants from both point and non-point sources, and being important for carbon sequestration and emissions (Finlayson *et al.*, 2005). The value of the world's wetlands are increasingly receiving due attention as they contribute to a healthy environment in many ways. Wetlands are an essential part of human civilization, meeting many crucial needs for life on earth such as drinking water, protein production, energy, fodder, biodiversity, flood storage, transport, recreation, and climate stabilizers. They also aid in improving water quality by filtering sediments and nutrients from surface water. Wetland functions are defined as the normal or characteristic activities that take place in wetland ecosystems or simply the things that wetlands do. Wetlands are one of the most productive of all ecosystems, and carry out critical regulatory functions of hydrological processes within watersheds (Banner *et al.*, 1988). Regulating water quality, water levels, flooding regimes, and nutrient and sedimentation levels are a few of these processes (Gregory *et al.*, 1991). As with any natural habitat, wetlands are important in supporting species diversity and have a complex of wetland values. Moreover, the pattern of seasonal variation of the wetland affects the bird population fluctuation (Imran and Mithas, 2009). Even small wetlands are extremely important to the conservation of biodiversity because they provide critical breeding habitat where dispersed populations can exchange genetic material, reducing the risks of extinction (Semlitsch and Brodie 1998).

However, unsustainable use of wetlands without reckoning of their assimilative capacity constitutes major threat to the conservation and management of these vital biodiversity rich areas. Thus, restricting the prospects of future generation to utilize the benefits of the ecosystem services provided by these wetlands. They are vulnerable to even small changes in their biotic and abiotic factors. In recent years, there has been concern over the continuous degradation of wetlands due to unplanned developmental activities. They are becoming extinct due to manifold reasons, including anthropogenic and natural processes. Burgeoning populations, intensified human activity, unplanned development, absence of management structures, lack of proper legislation, and lack of awareness about the vital role played by these ecosystems are the important causes that have contributed to their decline and loss.

The present review is aimed at providing the distribution of wetlands, the value of wetlands, the causes consequences of the loss of wetlands and current status of the management of wetland.

Distribution of Wetlands in India

India has a wealth of wetland ecosystems distributed across various eco-geographical regions that range from Himalayas to Deccan plateau. Varied topography and climatic regimes support and sustain diverse and unique wetland habitats in our country.

In India a total area of 40494 square km is classified as wetlands. This consists only 1.21 per cent of the total land surface. Natural wetlands in India consist of high altitude wetlands in Himalayas; flood plains of the major river 4 systems; saline and temporary wetlands of the arid and semi-arid regions; coastal wetlands such as lagoons, backwaters, estuaries, mangroves, swamps and coral reefs, and so on. In addition to these natural wetlands, a large number of man-made wetlands, which have resulted from the needs of irrigation, water supply, electricity, fisheries and flood control, are substantial in number. Wetlands in India occupy 58.2 million hectares, including areas under wet paddy cultivation (Directory of Indian Wetlands). These wetlands can be classified into different categories on the basis of their origin, vegetation, nutrient status and thermal characteristics. In India, out of an estimated 4.1 m ha (excluding irrigated agricultural lands, rivers, and streams) of wetlands, 1.5 m ha are natural, while 2.6 m ha are manmade. The coastal wetlands occupy an estimated 6,750 sq km, and are largely dominated by mangrove vegetation. The Wildlife Institute of India's survey reveals that they are disappearing at a rate of 2 per cent to 3 per cent every year. Most of the wetlands in India are directly or indirectly linked with major river systems such as the Ganga, the Cauvery, the Krishna, the Godavari and the Tapti.

Indian wetlands are grouped as:

Himalayan Wetlands

Ladakh and Zanskar: Pangong Tso, Tso Morari, Chantau, Noorichan, Chushul and Hanlay marshes.

Kashmir Valley: Dal, Anchar, Wular, Haigam, Malgam, Haukersar and Kranchu lakes.

Central Himalayas : Nainital, Bhimtal and Naukuchital.

Eastern Himalayas: Numerous wetlands in Sikkim, Assam, Arunachal Pradesh, Meghalaya, Nagaland and Manipur, Beels in the Brahmaputra and Barak valley.

Indo-Gangetic Wetlands

The Indo-Gangetic flood plain is the largest wetland system in India, extending from the river Indus in the west to Brahmaputra in the east. This includes the wetlands of the Himalayan terai and the Indo-Gangetic plains.

Coastal Wetlands

The vast intertidal areas, mangroves and lagoons along the 7500 kilometer long coastline in West Bengal, Orissa, Andhra Pradesh, Tamil Nadu, Kerala, Karnataka, Goa, Maharashtra and Gujarat. Mangrove forests of the Sunderbans of West Bengal and the Andaman and Nicobar Islands. Offshore coral reefs of the Gulf of Kutch, Gulf of Mannar, Lakshwadeep and Andaman and Nicobar Islands.

Wetland Values

The interaction of man with wetlands during the last few decades has been of concern largely due to the rapid population growth–accompanied by intensified industrial, commercial and residential development further leading to pollution of wetlands by domestic, industrial sewage, and agricultural run-offs as fertilizers, insecticides and feedlot wastes. The fact that wetland values are overlooked has resulted in threat to the source of these benefits.

Wetlands are often described as "kidneys of the landscape" (Mitsch & Gosselink 1986). Wetlands are among the most productive ecosystems. They directly or indirectly support millions of people and provide goods and services to them. Regional wetlands are integral parts of larger landscapes, their functions and values to the people in these landscapes; depend on both their extent and their location. Each wetland thus is ecologically unique. Various goods and services provided by wetlands are as follows:

- ☆ Support all life forms through extensive food webs,
- ☆ Habitat to aquatic flora and fauna, as well as numerous species of birds, including migratory species,
- ☆ Filtration of sediments and nutrients from surface water,
- ☆ Nutrients recycling,
- ☆ Water purification,
- ☆ Floods mitigation,
- ☆ Maintenance of stream flow,
- ☆ Ground water recharging,
- ☆ Provide drinking water, fish, fodder, fuel, etc.,
- ☆ Control rate of runoff in urban areas,
- ☆ Buffer shorelines against erosion,

- ☆ Comprise an important resource for sustainable tourism, recreation and cultural heritage,
- ☆ Stabilization of local climate,
- ☆ Source of livelihood to local people,
- ☆ Genetic reservoir for various species of plants (especially rice).

Wetland Losses: A Threat to Ecological Balance

Wetlands are one of the most threatened habitats of the world. Wetlands in India, as elsewhere are increasingly facing several anthropogenic pressures. The current loss rates in India can lead to serious consequences, where 74 per cent of the human population is rural (Anon. 1994) and many of these people are resource dependent. Coastal ecosystems are among the most productive yet highly threatened systems in the world. These ecosystems produce disproportionately more services relating to human well-being than most other systems, even those covering larger total areas, but are experiencing some of the most rapid degradation and loss. About 35 per cent of mangroves have been lost over the last two decades, driven primarily by aquaculture development, deforestation, and freshwater diversion. Some 20 per cent of coral reefs were lost and more than a further 20 per cent degraded in the last several decades of the twentieth century through overexploitation, destructive fishing practices, pollution and siltation and changes in storm frequency and intensity. The primary indirect drivers of degradation and loss of rivers, lakes, freshwater marshes, and other inland wetlands (including loss of species or reductions of populations in these systems) have been population growth and increasing economic development. The primary direct drivers of degradation and loss include infrastructure development, land conversion, water withdrawal, pollution, overharvesting and overexploitation, and the introduction of invasive alien species. Healthy wetlands are essential in India for sustainable food production and potable water availability for humans and livestock. They are also necessary for the continued existence of India's diverse populations of wildlife and plant species; a large number of endemic species are wetland dependent. Most problems pertaining to India's wetlands are related to human population. India contains 16 per cent of the world's population, and yet constitutes only 2.42 per cent of the earth's surface. Indian landscape has contained fewer and fewer natural wetlands over time. Restoration of these converted wetlands is quite difficult once these sites are occupied for non wetland uses. Hence, the demand for wetland products (*e.g.*, water, fish, wood, fiber, medicinal plants etc.) will increase with increase in population. Wetland loss refers to physical loss in the spatial extent or loss in the wetland function. The loss of one square km of wetlands in India will have much greater impacts than the loss of one km^2 of wetlands in low population areas of abundant wetlands (Foote Lee *et al.*, 1996).

The wetland loss in India can be divided into two broad groups namely, *acute and chronic losses*. The filling up of wet areas with soil constitutes acute loss whereas the gradual elimination of forest cover with subsequent erosion and sedimentation of the wetlands over many decades is termed as chronic loss.

Acute Wetland Losses

Direct Deforestation in Wetlands

There is removal of vegetation in the catchment leading to soil erosion and siltation. Mangrove vegetation are flood and salt tolerant and grow along the coasts and are valued for fish and shellfish, livestock fodder, fuel wood, building materials, local medicine, honey, bees wax and for extracting chemicals for tanning leather (Ahmad 1980). Alternative farming methods and fisheries production has replaced many mangrove areas and continues to pose threats. Eighty percent of India's 4240 square km of mangrove forests occur in the Sunderbans and the Andaman and Nicobar Islands (Anon. 1991). But most of the coastal mangroves are under severe pressure due to the economic demand on shrimps. Important ecosystem functions such as buffer zones against storm surges, nursery grounds and escape cover for commercially important fishery are lost. The shrimp farms also caused excessive withdrawal of freshwater and increased pollution load on water like increased lime, organic wastes, pesticides, chemicals and disease causing organisms. The greatest impacts were on the people directly dependent on the mangroves for natural materials, fish proteins and revenue. The ability of wetlands to trap sediments and slow water is reduced.

Agricultural Conversion

In the Indian subcontinent due to rice culture, there has been a loss in the spatial extent of wetlands. Rice farming is a wetland dependent activity and is developed in riparian zones, river deltas and savannah areas. Due to captured precipitation for fishpond aquaculture in the catchment areas and rice-farms occupying areas that are not wetlands, water is deprived to the downstream natural wetlands. Around 1.6 million hectares of freshwater are covered by freshwater fishponds in India. Rice-fields and fishponds come under wetlands, but they rarely function like natural wetlands. Of the estimated 58.2 million hectares of wetlands in India, 40.9 million hectares are under rice cultivation (Anon. 1993).

Inundation by Dammed Reservoirs

Presently, there are more than 1550 large reservoirs covering more than 1.45 million ha and more than 100000 small and medium reservoirs covering 1.1 million ha in India (Gopal 1994). By impounding the water, the hydrology of an area is significantly altered and allows for harnessing moving water as a source of energy. While the benefits of energy are well recognized, it also alters the ecosystem.

Hydrologic Alteration

Alteration in the hydrology can change the character, functions, values and the appearance of wetlands. The changes in hydrology include either the removal of water from wetlands or raising the land-surface elevation, such that it no longer floods. Canal dredging operations have been conducted in India from 1800s due to which 3044 km^2 of irrigated land has increased to 4550 km^2 in 1990 (Anon. 1994). Initial increase in the crop productivity has given way for reduced fertility and salt accumulations in soil due to irrigated farming of arid soils. India has 32,000 ha of peat-land remaining and drainage of these lands will lead to rapid subsidence of soil surface.

Chronic Wetland Losses

Introduced Species and Extinction of Native Biota

Wetlands in India support around 2400 species and subspecies of birds. But losses in habitat have threatened the diversity of these ecosystems (Mitchell & Gopal 1990). Introduction of exotic species like water hyacinth (*Eichornia crassipes*) and salvinia (*Salvinia molesta*) have threatened the wetlands and clogged the waterways competing with the native vegetation. In a recent attempt at prioritization of wetlands for conservation, Samant (1999) noted that as many as 700 potential wetlands do not have any data to prioritize. Many of these wetlands are threatened.

Groundwater Depletion

Recent estimate indicates that in rural India, about 6000 villages are without a source for drinking water due to the rapid depletion of ground water and draining of wetlands has depleted the ground water recharge.

Degradation of Water Quality

Over withdrawal of groundwater has led to salinization as the water quality is directly proportional to human population and its various activities. More than 50,000 small and large lakes are polluted to the point of being considered 'dead' (Chopra 1985). The major polluting factors are sewage, industrial pollution and agricultural runoff, which may contain pesticides, fertilizers and herbicides.

Wetlands and Climate Change

The role of wetland flux of carbon in the global carbon cycle is poorly understood. Wetlands may affect the atmospheric carbon cycle in four ways. *Firstly*, many wetlands especially boreal and tropical peat-lands have highly labile carbon and these wetlands may release carbon if water level is lowered or management practices results in oxidation of soils. *Secondly*, the entrance of carbon dioxide into a wetland system is via photosynthesis by wetland plants giving it the ability to alter its concentration in the atmosphere by sequestrating this carbon in the soil. *Thirdly*, wetlands are prone to trap carbon rich sediments from watershed sources and may also release dissolved carbon into adjacent ecosystem. This in turn affects both sequestration and emission rates of carbon. Lastly, wetlands are also known to contribute in the release of methane to the atmosphere even in the absence of climate change. Kasimir-Klemedtsson *et al.*, (1997) examined the conversion of bogs and fens to different cropping types that led to 23 fold increase in carbon dioxide equivalent emission. According to an estimation by Maltby *et al.*, (1993) when peat-lands are drained, the mineralization process starts immediately and results in emission of carbon dioxide ranging between 2.5 and 10t C2.5 and 10t C/ha/yr. Degradation of wetlands and disturbance of their anaerobic environment lead to a higher rate of decomposition of the large amount of carbon stored in them and thus release green house gases (GHGs) to the atmosphere. Therefore, protecting wetlands is a practical way of retaining the existing carbon reserves and thus avoiding emission of carbon dioxide and GHGs. Impacts of climate change on wetlands are still poorly understood. The diverse functions of wetlands make it more difficult to assess the relation between climate change and wetlands. The projected changes in climate are likely to affect the extent and nature of wetland functions. It even affects the role of wetlands as a sink of GHGs and reduces carbon storage and sequestration within them. It is uncertain if the conservation of wetland will be integrated into international trading schemes of emission as in Kyoto Protocol as of Forestry. Even trading of emission certificated may become an established pathway, and then mechanism can be applied to those wetlands with high carbon sequestration potential.

Wetland Management

Wetlands are not delineated under any specific administrative jurisdiction. Although some wetlands are protected after the formulation of the Wildlife Protection Act, the others are in grave danger of extinction. In India the conservation and wise use of wetlands falls within the mandate of the Central Ministry of Environment and Forests (MoEF). Some of the various central government agencies that may be indirectly making decisions which affect wetlands are: the Department of Fisheries, the Ministry of Agriculture, the Ministry of Water Resources, the Ministry of Surface Transport, the Ministry of Power, the Ministry of Tourism, the Department of Ocean Development, to name a few. Since land is a state subject (under the Constitution of the country), various state government agencies are also involved in decision making over wetlands (which are often equated to land). The numerous agencies (Government as well as private) involved in decision making on wetlands, make implementation of existing legal provisions for wetland conservation and wise use all the more difficult.

Protection Laws and Government Initiatives

Wetlands conservation in India is indirectly influenced by an array of policy and legislative measures (Parikh & Parikh 1999). Some of the key legislations are given below:

The Indian Fisheries Act – 1857, The Indian Forest Act – 1927, Wildlife (Protection) Act – 1972, Water (Prevention and Control of Pollution) Act – 1974, Territorial Water, Continental Shelf, Exclusive Economic Zone and other Marine Zones Act – 1976, Water (Prevention and Control of Pollution) Act – 1977, Maritime Zone of India (Regulation and fishing by foreign vessels) Act – 1980, Forest (Conservation Act) –1980, Environmental (Protection) Act – 1986, Coastal Zone Regulation Notification – 1991, Wildlife (Protection) Amendment Act – 1991, National Conservation Strategy and Policy, Statement on Environment and Development 1992, National Policy and Macro level Action Strategy on Biodiversity –1999.

India has set up 505 Wildlife Sanctuaries and 100 National Parks, 14 Biosphere Reserves, 6 Heritage Sites, Projects on Tiger conservation and Elephant conservation and Marine Turtles conservation with the objective of effective conservation of wetlands, and floral and faunal wealth in forest areas. India is an also signatory to the Ramsar Convention on Wetlands and the Convention of Biological Diversity. Apart from government regulation, development of better monitoring methods is needed to increase the knowledge of the physical and biological characteristics of each wetland resources, and to gain, from this knowledge, a better understanding of wetland dynamics and their controlling processes. India being one of the mega diverse nations of the world should strive to conserve the ecological character of these ecosystems along with the biodiversity of the flora and fauna associated with these ecosystems.

National Wetland Strategy

National wetland strategy should encompass

1. Conservation and collaborative management,
2. Prevention of loss and restoration, and
3. Sustainable management.

To achieve these strategies include the following:

Planning, Managing and Monitoring

Wetlands that come under the Protected area network have management plans but others do not. It is important for various stakeholders along with the local community and corporate sector to come together for an effective management plan. Active monitoring of these wetland systems over a period of time is essential.

Comprehensive Inventory

There has been no comprehensive inventory of all the Indian wetlands despite the efforts by the Ministry of Environment and Forests, Asian Wetland Bureau and World Wide Fund for Nature. The inventory should involve the flora, fauna, and biodiversity along with values. It should take into account the various stakeholders in the community too.

Legislation

There is no special legislation pertaining specially to these ecosystems so environment Impact Assessment needed for major development projects highlighting threats to wetlands need to be formulated.

Protection

The primary necessity today is to protect the existing wetlands. Of the many wetlands in India only around 68 wetlands are protected. But there are thousands of other wetlands that are biologically and economically important but have no legal status.

Research

There is a necessity for research in the formulation of national strategy to understand the dynamics of these ecosystems. The scientific knowledge will help the planners in understanding the economic values and benefits, which in turn will help in setting priorities and focusing the planning process.

Coordinated Approach

Since wetlands are common property with multi-purpose utility, their protection and management also need to be a common responsibility. An appropriate forum for resolving the conflict on wetland issues has to be set up. It is important for the ministries to allocate sufficient funds towards the conservation of these ecosystems.

Building Awareness

For achieving any sustainable success in the protection of these wetlands, awareness among the general public, educational and corporate institutions must be created. The policy makers, at various levels along with site managers need to be educated.

Use of Remote Sensing and GIS in Wetland Management

Remote sensing data in combination with Geographic Information System (GIS) are effective tools for wetland conservation and management. The application encompasses water resource assessment, hydrologic modeling, flood management, reservoir capacity surveys, assessment and monitoring of the environmental impacts of water resources project and water quality mapping and monitoring (Jonna 1999).

Monitoring of Irrigation and Cropping Pattern

Remote sensing data in association with the geographical information systems provides a cost and time-effective tool for identification, mapping, inventorying and monitoring of cropping pattern, crop production and condition, monitoring irrigation status and in the diagnosis of poorly performing irrigation patterns. These inventorying data are used as inputs for formulation of conservation and management plans for development of land and water resources.

Conclusion

Wetlands are a common property resource, it is an uphill task to protect or conserve the ecosystems unless; the principal stakeholders are involved in the process. Threats to wetland ecosystems comprise the increasing biotic and abiotic pressures and perils. Many of the wetlands are threatened. In India, so far as current status of wetland management is concerned, wetlands are not delineated under any specific administrative jurisdiction. The primary responsibility for the management of these ecosystems is in the hands of the Ministry of Environment and Forests. Although some wetlands are protected after the formulation of the Wildlife Protection Act, the others are in grave danger of extinction. Effective coordination between the different ministries, energy, industry, fisheries revenue, agriculture, transport and water resources, is essential for the protection of these ecosystems. The dynamic nature of wetlands necessitates the widespread and consistent use of satellite based remote sensors and low cost, affordable GIS tools for effective management and monitoring.

References

Ahmad N (1980). Some aspects of economic resources of Sunderbans mangrove forests of Bangladesh. pp. 50-51. In: P. Soepadmo (ed.) Mangrove Environment: Research and Management. Report on UNESCO.

Anonymous (1991). A Reference Annual. Research and Reference Division, Ministry of Information and Broadcasting, Govt. of India, Delhi.

Anonymous (1994). World Development Report. World Bank Development Report.

Asian Symposium, held at University of Malaya, Kuala Lumpur, Malaysia, 25-29 August 1980.

Banner A, Hebda RJ, Oswald ET, Pojar J, Trowbridge R (1988). Wetlands of Pacific Canada. In Wetlands of Canada, National Wetlands Working Group. Polyscience, Ottawa, pp. 306–346.

Chopra R (1985). The State of India's Environment. Ambassador Press, New Delhi.

Das S, Behera SC, Kar A, Narendra P, Guha S (1997). Hydro geomorphological mapping in ground water exploration using remotely sensed data – A case study in Keonjhar District, Orissa. *Journal of the Indian Society of Remote Sensing* 25: 247-250.

Finlayson, C.M., R. D'Cruz, N. Davidson, J. Alder, S. Cork, R. de Groot, C. Leveque, G.R. Milton, G. Peterson, D. Pritchard, B.D. Ratner, W.V. Reid, C. Revenga, M. Rivera, F. Schutyser, M. Siebentritt, M. Stuip, R. Tharme, S. Butchart, E. Dieme-Amting, H. Gitay, S. Raaymakers and D. Taylor, Eds., (2005). Ecosystems and Human Well-being: Wetlands and Water Synthesis. Island Press, Washington, District of Columbia, 80 pp.

Foote Lee S, Pandey, Krogman NT (1996). Processes of wetland loss in India. Environmental Conservation 23: 45-54.

Gitelson A, Garbuzov G, Szilagyi F, Mittenzwey KH, Karnielli AQ, Kaiser A (1993). Quantitative remote sensing methods for real-time monitoring of inland water quality. *International Journal of Remote Sensing* 14: 1269-1295.

Gopal B (1994). Conservation of inland waters in India: an overview. *Verhandlungen der Internationalen Vereinigung fur Theorestische und Angewandie Limnologie* 25: 2492-2497.

Gregory SV, FJ Swanson, WA McKee, Kenneth WC (1991). An ecosystem perspective of riparian zones. *Bioscience* 41: 540–550.

Imran A. Dar, Dar Mithas (2009). Seasonal variation of Bird Population in Shallabug Wetland Kashmir, India: *Journal of Wetland Ecology*, Vol. 2: 19-33.

Jonna S (1999). Remote sensing applications to water resources: Retrospective and Perspective. pp. 368- 377. In: S. Adiga (ed.). Proceedings of ISRS National Symposium on Remote Sensing Applications for Natural Resources. Dehradun.

Kasimir-Klemedtsson L, Berglund K, Martikainen P, Silvola J, Oenema O (1997). Greenhouse gas emission of methane from farmed organic soils: a review. *Soil Use Manage.*, 13: 245-250.

Maltby E, Immirzi CP (1993). Carbon dynamics in peatlands and the other wetlands soils: regional and global perspectives. *Chemosphere*, 27: 999-1023.

Mitchell S, Gopal B (1990). Invasion of tropical freshwater by alien species. pp. 139-154. In: P. S. Ramakrishnan (ed.) Ecology of Biological Invasion in the Tropics.

Mitsch WI, Gosselink IG. (1986). Wetlands. Van Nostrand Reinhold, New York.

Mohan S, Shrestha MN (2000). A GIS based integrated model for assessment of hydrological changes due to land-use modifications. pp. 27-29. In: T.V. Ramchandra (ed.) Symposium on Restoration of Lakes and Wetlands, November 2000, Indian Institute of Sciences, Bangalore.

Parikh J, Parikh K (1999). Sustainable Wetland. Environmental Governance – 2, Indira Gandhi Institute of Development Research, Mumbai.

Reid WV, Mooney HA, Cropper A, Capistrano D, Carpenter SR, Chopra K, Dasgupta P, Dietz T, Duraiappah AK, Hassan R, Kasperson R, Leemans R, May RM, McMichael AJ, Pingali P, Samper C, Scholes R, Watson RT, Zakri AH, Shidong Z, Ash NJ, Bennett E, Kumar P, Lee MJ, Raudsepp C, Hearne H. Simons, Thonell J, Zurek, MB (2005). Ecosystems and Human Well-being: Synthesis. Island Press, Washington, District of Columbia, 155 pp.

Sasmal SK, Raju PLN (1996). Monitoring suspended load in estuarine waters of Hooghly with satellite data using PC based GIS environment.

Semlitsch RD, Brodie RD (1998). Are small, isolated wetlands expendable? *Conservation Biology* 12: 1129–1133.

Seshamani R, Alex TK, Jain YK (1994). An airborne sensor for primary productivity and related parameters of coastal waters and large water bodies. *International Journal of Remote Sensing* 15:1101-1108.

Chapter 6

Microbial Diversity: Potential and Challenges

☆ *Sadhana Pandey, C.J. Mehta*
& P. Kulshrestha

There has been a remarkable resurgence in the interest in diversity of life forms existing on this planet–"The earth". The signatories' nations of Earth summit 1992 are committed to explore, identify and conserve diverse forms of life. India has taken strides in meeting the follow up agenda -21, Objectives of Earth summit. Some initiative in the form- Biodiversity Act-2000, National Biodiversity strategy and Plan are taking care of life form occurring on our motherland. However, this is not adequate and much more is required in general and microorganisms in particular.

Diversity is composed of two elements- richness and evenness. So the highest diversity occurs in communities with many different species present (richness) in relatively equal abundance (evenness). According to Huston 1994, the richness and evenness of microbial community reflects selective pressure that shape diversity within communities.

It is unfortunate that microorganisms have not received proper attention and have largely been ignored in international and national policies on biodiversity programs. Microbial diversity includes Bacteria, Fungi, Protozoa and Unicellular algae and constitutes the most extraordinary reservoir of life in biosphere.

Microbes have been evolving for nearly 4 billion years and were the only form of life on earth. During this long period of time all basic biochemistry of life evolved in these forms. It is estimated that 50 per cent of living protoplasm on this planet is microbial (Stanley 2002). They represent the richest repertoire of molecular and chemical diversity in nature. They underlie basic ecosystem process as such as biogeochemical cycles and food chain as well as maintain vital and elegant relationship between themselves in higher organisms. Without these tiny organisms all life on earth would cease.

Microorganisms should not be counted as additional element to the biological diversity on the earth. They are inseparable part of life and man has long exploited this metabolic wealth. There are various fields in which microbes play very important role.

New Dimensions in Agriculture

Microbes have many functional roles in any ecosystem. They are decomposers on one hand and large sinks of nutrients on the other. They prevent nutrient leaching and their loss from the soil. Mycorrhizal association support carbon fixation in symbiotic association with trees (Smith and Read 1997). Other fungal structures such as rhizomorphs retain nutrients, thus preventing nutrient loss from soil. Some fungi also exude polysaccharide similar to exudates of roots and these organic glues are important in creating and establishing soil micro aggregates and soil micropores (Perry *et al.*, 1989). These soil structures contribute to soil aeration which is critical for overall maintenance of soil health and productivity (Molina *et al.*, 2001).

Microbes have the potential to be used as bio-control agent. Bacteria and some fungi have been screened and some are being used commercially against various plant diseases. Since 1980's, Microbial Pest Control Agents (MPCAs) have been gaining prominence because biotechnology has offered newer options in the selective manipulation and application of microbial pesticides. In South Africa coffee bean borer has been successfully controlled by conidia of *Beauveria bassiana* (Murin 1996). Another success story has been presented by Brazil where *Metarhizium anisopliae* controlled population of sap sucking bug in rice field has been reported (Moldenk 2003).

Neutraceuticals and Pharmaceuticals

We have long exploited the metabolic pathways of these microorganisms to produce food and health applications. Solid substrate technology is now available which uses variety of agricultural and other waste- to produce food and food supplements.

Edible mushrooms are much more prolific and have higher biological efficiency when compared with many plant products. Nutritional status of some mushroom (*Cantharellus* sp., *Lentinus cladopus, Pleurotus floucla, Tricholoms georgii*) have been found much superior and emphasis is being given for recovery and conservation of green germplasm for future applications. The annual yield of protein by button mushrooms, *Agaricus bisporus* is 22 tonnes/ha. In contrast, the corresponding yield of protein for most plants is 1-2 tonnes/ha/yr (Yang 1988, Zang 2000). These mushrooms besides being nutritive also endowed with medicinal attributes including anticancer immuno enhancing, anti diabetic and hypolepidemic activities. (Chang & Buswell 1996).

For many countries food production is the key to development but is often restricted by available energy resources and environment problems such as drought. Research in the field of microbial biotechnology has yielded variety of Microbial Biomass Protein (MBP) or mycoprotein. Quorn (mycoprotein) produced from *Fusarium grameniarum* is one of the best examples of MBP. This mycoprotein has an annual market of 25 million sterling pounds.

Earlier studies suggested that some live bacteria like *Lactobacellus vulgaricus* and *Streptococcus thermophilic,* in the form of yogurt, have beneficial effects on human health. Today this has been scientifically validated and a range of commercial products are available in the market as Probiotics. These probiotic food items, now recommended by doctors are supplemented with medicines (Da Silva and Taguchi 1987).

To illustrate the extraordinary diversity in terms of secondary metabolites, it is worthwhile that approx. 3500 antibiotic secondary metabolites have been recognized from the genus *Streptomyces* alone.

The bioactive molecule producing capability of fungi is also enormous. Nielsen and Smedsgaard (2003) have presented at least of 474 fungal metabolites or *Exatrolites* (A term used for an outwardly directed chemically differentiable product of living organisms). In recent years advances in molecular biology automation and computer science have changed the direction of screening for novel compounds including enzymes, enzyme inhibitors, antihelminthics, antitumour agent, insecticides, vitamins, immunosuppresents and immuno-modulators. These compounds represent only a small portion of what is likely present in nature.

Energy

Anaerobic fermentation technology employs number of bacteria for the production of biogas and fertilizers. This technology has been very popular. It is eco-friendly which uses domestic, agricultural and other organic wastes.

Bioremediation

The microorganisms are also being used in monitoring and managing pollutants in the environment. Bioremediation is the process by which contamination in soil/water is cured by microbial biogeochemical process. It is an economical, versatile and environmental friendly, efficient treatment technology and is emerging technology of environmental restoration. It is self sustaining and inexpensive (Watanabe *et al.*, 2002). It has been established that contaminated environment harbors a wide range of unidentified pollutant degrading microorganisms. Efforts are being made to characterize bacterial communities and their responses to pollutants, and also to identify the genes involving particular degradation process (Margesin *et al.*, 2003).

Microbes are also used for bioleaching of metals, monitoring of pollutants, cleaning up of oils spills, waste water treatment and also tools for medical research.

Conclusion

The biological diversity of Indian subcontinent is one of the richest in the world owing to its vast geographic area, varied topography and several bio-geographical regions. Microorganisms share more than 18 per cent of total Indian biodiversity. Thus it can be concluded that microorganisms are one of the most important components of ecosystem and bio-resources for both basic and applied research. New emerging molecular methods and computer methods are now available to deal with numerous potential characters. These factors permit new and exciting insight, never possible previously, in the systematics of fungi. On the other hand, their enormous commercial potential may contribute to sustainable long term benefit to human beings. Countries like India must realize the value of their microbial resources and should promote microbial inventories and establish germplasm collection centers. Establishment of regional base centers, in different regions, for germplasm collection and *ex situ* conservation is recommended.

References

Chang ST and Buswell JA (1996). World J. Microbiology and Biotechnology 12: 473-476.

Da Silva EJ and Taguchi H (1987). An international network exercise: the MIRCEN programme. In Microbial Technology in the Developing World, Oxford Uni. Press; Oxford,UK, pp. 313-335.

Huston MA (1994). Biological Diversity, Cambridge University Press, UK.

Margesin R, Labbe D, Schinner F, Greer CW, Whyte LG (2003). Appl. Environ. Microbiol. 69: 3985-3992.

Moldenke AR (2003). Curr. Sci. 84(5): 617-618.

Molina R, Massicotte HB and Trappe JM (1992). In Mycorrhizal Functioning: An integrative Plant Fungal Process. Chapman and Hall, New York. 357-423.

Molina R, Pilz D, Smith J, Dunham S, Dreisbach T, O'Dell T and Castellans M (2001). Conservation and Management of forest fungi in the pacific North Western United States and integrated ecosystem approach. In Fungal conservation: Issues and Solutions, Cambridge Uni. Press, Cambridge, UK. pp.19-63.

Nielsen KF and Smedsgaard J (2003). J. Chromatography. 1002: 11-136.

Perry DA, Amaranthus MP, Borchers JG, Borchers SL and Brainerd RE (1989). Bioscience 39: 230-237.

Smith SE and Read DJ (1997). Mycorrhizal Symbiosis, Academic Press, London.

Stanley J (2002). Biodiversity of Microbial life. Wiley-Liss, New York

Trinci APJ (1992). Mycol. Res. 96: 1-13.

Watanabe K, Futamata H, Harayama S (2002). Anton. Van Leeuwen. 81: 655-663.

Yang XM (1988). Introduction: In Methods for cultivation of Edible Mushrooms in China, Chinese Agri. Press, Beijing. 19.

Zang GY (2000). Chinese Edible Fungi. 19: 2.

Chapter 7

Insect Biodiversity and Conservation of Insects

☆ *Mukulita Upadhyay*

Biodiversity is one of the important milestones of sustainable development and is known to be the biological wealth of any nation. India is among the twelve megabiodiversity countries of the world. Insects comprises of largest group of organisms and constitute an important aspect of biodiversity. Insects are involved in providing various services to the ecosystem such as pollination, silk, honey and lac production. 80 per cent of the insects are endemic in India. Over 1 million species of insects have been described, but current estimates of total insect diversity vary from 5-80 million species of insects. Beetles make up 40 per cent of described insect species Flies (Diptera) and Hymenoptera (Wasps, Bees and Ants) could be diverse or more so. Five orders of insects stand out in their levels of species richness : Hymenoptera, Diptera, Coleoptera, Lepidoptera and Hemiptera.

Phylum Arthropoda is the largest of all phyla as it ranks high in the scale of animal life. Arthropoda has a vast assemblage of metamerically segmented animals with jointed appendages and chitinous exoskeleton such as scorpions, spiders, centipede, millipede and their allies.

Arthropods are considered to be most successful among all the inhabitants of the terrestrial ecosystem because of their adaptive diversity and this has made them possible to survive in virtually every habitat. Arthropods occupy the first place in the animal kingdom considering few criterias such as the known species and individual, their distribution, their feeding patterns, their tolerance level to kinds of habitat they invade and their ability to defend against enemies. They display complex behaviour patterns and are highly social. Arthropods are the only major invertebrates adapted to live on dry lands and insects invade both land and aerial environment. As a result of such invasion they occupy every conceivable ecological niche. They have inhabited diverse habitats all over the globe. They inhabit sulphur springs, hot spring freshwater, lakes, streams, ponds, deserts, underground, as ecto and endo parasite and as pests. They crawl, fly, swim, hop and just set still. The diversity of forms of these animals seems infinite but above all, they display a basic body plan. They, hence, evolved social organization with a well marked division of labour among members of different castes.

Diversity

Arthopods belong to an ancient stock and were well diversified in the past. Today, they are the most dominant animals on the earth, considering the number of species.

One estimate indicates that arthropods have 1,170,000 described species, and account for over 80 per cent of all known living animal species. Another study estimates that there are between 5 to 10 million arthropod species, both described and yet to be described. Estimating the total number of living species is extremely difficult because if often depends on a series of assumptions in order to scale up from counts at specific locations to estimates for the whole world.

Arthropods are important members of marine, freshwater, land and air ecosystems, and are one of only two major animal groups that have adapted to life in dry environments; the other is amniotes, whose living members are reptiles, birds and mammals. One arthropod sub-group, insecta, is the most species-rich member of all ecological quilds (ways of making a living) in land and fresh-water environments. The lightest insects weigh less than 25 micrograms, while the heaviest weigh over 70 grams. Some living crustaceans are much larger, for example the legs of the Japanese spider crab may span up to 4 metres. All these arthropods exhibit great diversity in the form of their habitats, body structure and forms, economic importance, social and religious beliefs.

There are millions of distinct species of arthropods including all insects, crustaceans, millipede, centipede, spiders, and a host of other animals, all united by having same body plan *i.e.* jointed legs and Chitinous exoskeleton. They are by far, the most numerous and most diverse of all creatures on Earth. They make up approx. 1.6 million of estimated 1.8 to 1.9 million described species. They are by far planet's most important group of animals in terms of their importance as pollinators to medical role (*e.g.* disease vectors and parasites) to biological control of introduced species, to toxicology and biopharmaceuticals.

Insect Diversity: Some Important Orders

The diversity among different orders is characterized on the basis of mouth part structure, number and structure of wings and type of metamorphosis.

Odonate

Dragon flies and damsel flies. They are characterized by lack of wing flexibility such that their wings are held spread out from their bodies at all time (except damselflies).They are active predators, primarily of aquatic habitats.

Lepedoptera

Butterflies & Moths. These insects have enlarged wings that are covered with scales. Most Lepidopteron have mouth parts organized as long coilable proboscis through which food is ingested.

After beetles, second largest subgroup of insect is butterflies and moths with about 175,000 species. They are known as Lepidoptera, a name that means "scaly winged" because their large papery wings are covered with tiny scales Lepidoptera show great diversity. Clearwings are day flying moths. They look more like butterflies with their bright colours & slim bodies. The hawk moths are fast and powerful fliers with narrow V shaped wings. The peacock butterfly has large patches on its wings that look like eyes and are called eyespots. These wings are helpful to frighten away predators.

Orthoptera

Examples–Crickets and Grasshoppers. Medium to large sized insects, members of this order have enlarged hindlegs for jumping, mandibulate mouth parts, and, pleated wings that can fold like a fan when not in use. The diversity among this order is exhibited by comoflanged bodies, coloured and patterned to blend in with their surroundings. In few members of this order the bodies are modified to resemble pieces of vegetation for example stick insect.

Coleoptera

Beetles, members of this order are characterized by sclerotized forewings (elytra). This is the largest order of insects with over 2,50,000 described species. There is a great diversity in structure and life style among its members. Most are terrestrial herbivores but there are also predatory species. Beetles are the most wide spread of all insects.

They live in almost every habitat and region of our planet except the sea. They are not only incredibly numerous, they are also extremely varied. Green tiger beetles are truly the tigers of insect world. They are active hunters. One kind of green ground beetle, the bombardier can squirt a spray of stinging chemicals from its rear end at attackers.

Not only Beetles but their smaller cousins weevils also constitute the largest subgroup of insects.

Weevils form largest subgroup of beetles with more than 60,000 species. They feed on flowers seeds & fruits.

Dung, scarab and minotaur bettles feed on detritus – animal dropping & bits of dead animals & plants and help to recycle minerals and nutrients.

Hemiptera

True Bugs, mouthparts are modified into a piercing structure that can probe plant or animal tissue and suck nutrients out. The wings are generally folded flat over the abdomen, and the basal portion of the wings is thickened or leathery.

Homoptera

Leaf hoppers, aphids, spittlebugs, whiteflies, scale insects, mealybugs, cicadas. These animals are herbivores that feed on the photosynthate in plant phloem. Their mouthparts are modified into a needle-like tube that pierces stems and probes to the phloem layer. Their membranous wings are often held roof-like over their abdomen. This order contains a large number of common plant pests, familiar to gardeners.

Diptera

Flies and mosquitoes. This order is characterized by only one pair of flying wings in adults. The hind wings are transformed into halteres, gyroscopic balancing organs that enhance rapid and efficient flight. Life styles of dipterans vary; most are saprophytic, living off of decaying material, while others are efficient pollinators.

Hymenoptera

Bees, Wasps and Ants. Members of this order have stiff, membranous wings. All members also possess well developed mandibles that have been adapted to a wide array of tasks. This group contains many of the more agriculturally beneficial insects, including pollinators, predators and plant-eating insects.

Conservation of Insects

Insects are cosmopolitan in distribution. Insect conservation is not a new idea it has been around for well over a century but now the insects are coming into their own and the importance of insect conservation is more and more being accepted by the people.

There are however, many reasons for this on a general level. The diversity of life on this planet is part of what makes it so great to live on. Reduce the diversity and you reduce the pleasure of every life. Another way to look at it is to realize that the perception of diversity and variety in the world around us is important for our mental and spiritual health.

We human beings get all sorts of generally unnoticed benefits from insects,–the aesthetic appreciation of their beauty gives us pleasure, the study of insects has been of invaluable in helping us come to understand the world we live in, and insects have and still do play a very important role in medical research and the curing of many diseases.

No aspect of your life is unaffected by insects in some way. As you would expect of an animal group that contains over 3/4 of the species recorded on the planet, the economic importance of insects is immense. It is true that some insects cost mankind money and resources, but these are only a very small percent of the total, and without them much of modern science would not exist. However, the beneficial aspects of insects far outweigh their negative costs.

The honey industry is supplying jobs for thousands of people and generating 1000s of millions of pounds of revenue every year. Without insect pollinators many of the plants that we use for food, construction, and for decoration could not exist.

It is true to say that without insects most of the terrestrial life forms on this planet would slowly disappear. Insects are a part of nearly every food chain, either directly as food for other insects, fishes, amphibians, reptiles, birds, mammals and other arthropods, or indirectly as agents in the endless recycling of nutrients in the soil. Insects and mites are extremely important in helping microbes break down dung, dead plant and dead animal matter in the soil and leaf litter layer, so that the nutrients that plants need to grow can be released into the soil. Though these insects are often small their importance is great and without them this important process would be at least, 10 times slower than it is with their help.

It is very difficult to conserve insects because it the functional level at is different for every species.

This is because to conserve insects we need to make sure everything they need to live is available for them when they need it. This is often more complicated than it seems because many insects use one habitat as a food source in the juvenile form, another as a food source as an adult while maturing, another for meeting mates and laying eggs and perhaps still another as a place to over winter.

'Insects Conservation' often involves 'Habitat conservation' instead. This works on the hope that if we preserve enough of each sort of habitat this will allow all the insects to survive as well. This is true to a certain extent but eventually we need to know all the requirements of each species of insect, if we wish to be sure of conserving them. As there are a million known species of insect on the planet, and perhaps as many as 30 million yet to be discovered, we can see there is a lot of work still to be done.

There are two approaches to the conservation of insects–either human set aside large portion of land using "Wilderness preservation" as the motive or confronting the particular processes that effect the charismatic vertebrates. Single species conservation is said to preserve many other species indirectly. This is referred as *umbrella effect*. Flagship species like butterflies, colorful beetles can expand public awareness and financial contribution for conservation efforts.

Migratory species such as Monarch butterfly (*Danaus plereppus*) is in tremendous need for conservation methods. One species requires several habitat locations for different periods.

Queen Alexandras Birdwing (*Oruthoptera alexandra*) is restricted to a very specific area due to specificity in their diet. The developing countries like Papua New Guinea have worked a lot for the conservation for such a valuable butterfly.

Western Ghats is one of the hotspots of biodiversity and needs urgent attention for conservation because of high degree of butterfly endemism and the grave threats it has. Forest insects should be collected, conserved and documented. Fungal pathogens should be employed for sustainable management of insect pests.

Mosquitoes are controlled principally by the use of synthetic insecticides which has adverse effects. Therefore, their is need for development of selective mosquito control alternatives. Plants and plant essential oils are relatively non-toxic therefore they are exempted from toxicity data requirements in many countries. They have been exploited as mosquito control agents.

Adult beetles is an indicator species of forest ecosystem and play an important role in litter degradation and nitrogen recycling mechanism at the forest floor. Therefore, methods should be employed to conserve beetles.

There is immense need for developing *in situ* and *ex situ* conservation of insect fauna and above all nature education and interpretation programmes should be promoted.

References

Foottit RG, Adler PH (2009). Insect biodiversity: science and society, pp. 290.

Gullan PJ, Cranston PS (2005). The insects: an outline of entomology, pp. 20.

Hawksworth DL , Bull AT (2006). Arthropod diversity and conservation.

Hosetti BB, Venkateshwarlu M (2001). Trends in Wildlife Biodiversity Conservation and Management.

Ignacimuthu S, Jayaraj S (2006). Biodiversity and insect pest management.

New TR (2010). Beetles in Conservation.

Watt AD , Stork NE , Hunter MD (1997). Forests and Insects.

Chapter 8

Ecological and Economical Impact of Biodiesel Crops Cultivation in Chambal Region

☆ *J.K. Mishra*

India needs availability of energy at genuine cost for economic development of the people. Uncertain supplies and fluctuation in oil prices in international market need a search for renewable, safe, non polluting sources of fuel. The so called Biofuels are at last becoming available alternative to gasoline and diesel. The economic condition of the farmers of Chambal region is very poor and their earning is based on agriculture production. Due to the formation and expansion of ravines in Chambal region, the area of agriculture land had been reduced to a great extent. The ravines were developed on both the sides of Chambal and its associates rivers *i.e.* Kawari, Assan, Sank, Parwati, Sindh Vaishali & Pahul upto 2 to 5 km. The total area of ravines of Chambal region was recorded 3.10 lacs hectare.

The climatic factors including soil, rainfall, temperature, irrigation facilities etc are not favorable for the sufficient crop yield. That is why the per capita income of farmers is not satisfactory. Majority of the farmer population is unable to fulfill their stomach and daily needs of life. This economic disparity had created the problem of migration from villages to cities, class struggle and social unrest amongst the people of this region. The Robber's problem is one of the burning example of social unrest. Hence there is an urgent need to reduce the economic disparity for making the life of the people economically comfortable.

Therefore the Central Govt. recently has encouraged the farmers for growing more and more biodiesel crops like Jatropha, Karanj, Neem and Mahua on the baren land of Chambal ravines. The non edible oil bearing trees have potential of producing Biodiesel/Bio fuel. The exploitation of these resources in arid and semi arid region including Chambal ravines will of vital importance for the farmers of developing country like India. The cultivation of Biodiesel crops will not only increase the per capita income of the poor farmers but this will also help to increase the tree cover area of the Chambal ravines, reduce desertification, soil, erosion, droughts and floods in, different regions of the country. Hence the cultivation of Biodiesel crop like *Jatropha* will prove as a boon for the farmers of the Chambal region from ecological and economical point of view.

Botanical Description of *Jatropha curcus* L.

Jatropha curcus L. belongs to the family Euphorbiaceae of dicot plants. It is native of tropical America. The plant is deciduous shrub or small tree having up to five meter height, young shoots glandular, tomentose base, leaves green alternate, orbicular, cordate having multi costate, reticulate venation, 3 to 5 palmately lobed. Flowers are yellowish green terminal and sub-terminal. Male flowers are unisexual, staminate, pentamarous while female flowers are tricarpellary having trifid stigma. Fruit is regma-splitting in to three cocci. The plant flowers nearly 10 to 12 months after planting, preferably during September – October. The fruit matures during March and April. The mature fruits are harvested, seeds are separated and used for oil extraction.

In India it is believed that *Jatropha* had been introduced by Portugase navigaters in the 16th Century. *Jatropha* is quick growing and has a remarkable adaptability to variety of climate and soils. It is found widely in waste lands, marginal land and dry exposed slopes of lower hills up to 1000 M. elevations. This crop is gaining momentum because this crop has been called as future fuel. Seeds of *Jatropha* contain 30 to 35 per cent oil by weight of air dried seed. This oil can be used as Biodiesel after trans esterification.

Soil and Climate for *Jatropha* Cultivation

Jatropha can thrive in tropical and sub-tropical climate excepting in places experiencing frost. It thrives even in alkaline, saline, acidic soils, ravines and degraded land, waste land and marginal land. The soil and climatic conditions of Chambal ravines are favourable for *Jatropha* cultivation.

Propagation of *Jatropha*

It is generally propagated by cutting. The one year old shoots are selected for propagation of cuttings. Thick strong shoots of 20 to 25 cm. long with 4 to 5 buds, preferably taken from the middle of the branch are best suited material for propagation as they give nearly 82 to 90 per cent rooting. The cuttings are planted closely with spacing of 15–20 cm. The bed are watered regularly, cutting will be ready for planting within 6 to 8 months.

Biodiesel Production from *Jatropha* Seed Oil at National and International Level

Traditionally, raw vegetable oil produced from *Jatropha* and several other crops has been used as such without any modification as fuel in rural areas for lighting purpose. It was also used for running tractors. Recently in France which is a leading producer of Biodiesel, commercial diesel contains upto 5 per cent of Biodiesel, whereas in USA blends upto 20 per cent are used. Biodiesel is a vegetable processed fuel which resembles to diesel fuel. Chemically, biodiesel is a methyl or ethyal esters of fatty acids made from virgin or used vegetable oils and animal fats. Biodiesel are prepared by trans esterification, also called as Alcoholysis. In this process the displacement of alcohol from an ester by another alcohol takes place. Biodiesel is ecofriendly with clean burning renewal fuel. It requires no engine modification and increases engine life. It is Biodegradble and non-toxic. According to a survey report of the year 2000. The Biodiesel sector of France had produced 328000 Mt.; Germany –

246300 Mt., Italy-78000 Mt., Austria- 27600 Mt. and Belgium produced – 20000 Mt. Biodiesel.

In India Coimbatore district had taken a lead in the mass cultivation and production of *Jatropha* based Biodiesel. The district rural development agency plans to setup a Bio-diesel plant at a cost of Rs 7.5 lacs under SGRY scheme in Annaikatty. The tribal self, help group had already pioneered by planting about 7.00 lacs samplings in 585 Acres in area like Thonda Muthur, Kariamandai, Jambukandi Tribal Village in Annaikatty. It is estimated that 3 to 4 kg of seeds can be harvested from a *Jatropha* tree annually. The expected cost of 1 Kg. seeds is about Rs. 5. From 4 Kg. seeds we can get 1 Kg. oil and 3 Kg. oil cake.

Many private companies, NGO's and DRDO started purchasing land specially in Gujarat, Orissa and Tamil Nadu for establishing *Jatropha* piantation.

Ecological and Economical Significance of *Jatropha* Cultivation

1. *Jatropha* seed oil based diesel offer lesser damage to the environment as against fossil fuel.
2. *Jatropha* cultivation will help to increase tree cover area and reduce desertification, soil erosion, in floods in different regions of our country including Chambal ravines.
3. Biodiesel crop also play an important role in CO_2 cycle of the earth since they assimilates CO_2.
4. There is a lot of employment opportunities for unskilled, semi-skilled and qualified labours and this will solve the migration problem of the villagers.
5. Oil cakes can be used for Bio-gas production.
6. Bio-fuels/diesel are in great demand in industries like thin film coating on metals, printing, ink for newspapers.
7. It will help in increasing the per capita income of the rural farmers which will improve their economic conditions.
8. Biodiesel crop cultivation will check further expansion of Chambal ravines.
9. It will reduce import oil bill and the make the country economically sound by cutting down the foreign expenditure.

10. It will provide the opportunity to the rural women to start cottage industries such as soap, perfume, oil and insecticide production.
11. Biodiesel is an environmentally clean fuel and economically cheaper for the users. It has no sulphur and aromatics, hence, it will also bring down the environmental pollution level.

Thus it can be said that in a holistic perspective *Jatropha*/ Biodiesel crops cultivation seems to be an ideal one from ecological and economical point of view in Chambal ravines.

References

Germplasm Resources Information Network. United States Department of Agriculture. 2007-10- 05. http://www.ars-grin.gov/cgi-bin/npgs/html/genus.pl?6189.

Achten WMJ, Mathijs E, Verchot L, Singh VP, Aerts R, Muys B (2007). Jatropha biodiesel fueling sustainability?. Biofuels, Bioproducts and Biorefining 1(4), 283-291.

Achten WMJ, Verchot L, Franken YJ, Mathijs E, Singh VP, Aerts R, Muys B (2008). Jatropha bio-diesel production and use. (a literature review) Biomass and Bioenergy 32(12), 1063-1084.

Fairless D. (2007). "Biofuel: The little shrub that could–maybe". *Nature* 449 (7163): 652–655.

http://www.drugsandpoisons.com/2008/01/lectins-peas-and-beans-gone-bad.html

MacIntyre, Ben (2007-07-08). "Poison plant could help to cure the planet". London: The Times. http://www.timesonline.co.uk/tol/news/world/article2155351.ece.

Chapter 9

Effect of Biodiversity Loss on Human Health

☆ *Rajeev Kumar Bhadkariya, Ashok Kumar, M.C. Pathak and B.R. Singh*

Health is our most basic right. Biodiversity is the foundation for the human health. Almost all medicines were derived from biological resources. Even today they remain vital and as much as 67–70 per cent of modern medicines are derived from natural products. In developing countries, a large majority of the people rely on traditional medicines for their primary health care, most of which involve the use of plant extracts. At least 50 per cent of the pharmaceutical compounds on the US market are derived from compound found in plants, animals and microorganisms, while 80 per cent world population depends on the medicine from nature (Behra *et al.*, 2008 and Marvier, 2007).

Environment change is increasingly driven by human activities which are resulting in global warming, habitat destruction, pollution, overharvesting and introduction of exotic species. These changes have caused decline in biodiversity at regional and global scales and have had major impacts on human health through direct and indirect impacts on infectious diseases, nutritional and contaminants (Rapport *et al.*, 2004).

Infectious Diseases and Biodiversity

Human activities are disturbing both the structure and functions of ecosystems and altering native biodiversity. Such disturbances reduce the abundance of some organisms, cause population growth in others, modify the interactions among organisms, and alter the interactions between organisms and their physical and chemical environments. Patterns of infectious diseases are sensitive to these disturbances. Major processes affecting infectious disease reservoirs and transmission include, deforestation; land-use change; water management *e.g.* through dam construction, irrigation, uncontrolled urbanization or urban sprawl; resistance to pesticide chemicals used to control certain disease vectors; climate variability and change; migration and international travel and trade; and the accidental or intentional human introduction of pathogens (WHO).

Some of the diseases are as follows, due to loss of the biodiversity:

Lyme Disease

Recent studies have demonstrated that suburban sprawl leading to forest fragmentation could increase the prevalence of ixodid ticks infected with *Borrelia burgdorferi*, the spirochaete bacterium that causes Lyme disease. Lyme disease is the most common tick-borne illness in North America and Europe. It feeds on the blood of animals and humans and can harbor the bacteria and spread it when feeding. Empirical studies confirm that disease risk is significantly higher in areas of low vertebrate diversity, such as small forests (less than 2 hectares) (Allan *et al.*, 2003) and highly fragmented landscapes (Brownstein *et al.*, 2005).

Malaria

Pongsiri and Roman's team, in 2006, in Amazonian Peru, demonstrated that malaria transmission can rise in response to deforestation. It appears that loss of the structural diversity provided by trees led to higher density of *Anopheles darlingi* mosquitoes, a potent transmitter of malaria, as well as to higher biting rates (Vittor *et al.*, 2006).

Yasuoka *et al.*, 2007 reported that changes in plant diversity–particularly through habitat alteration, fragmentation, and deforestation–can increase the risk of malaria transmission through effects on mosquito survival, density, and distribution. Another study demonstrated that deforestation can also increase transmission by raising surface-water availability and creating new breeding sites for some *Anopheles* mosquitoes (Walsh *et al.*, 1993).

Schistosomiasis

The loss of predators can cause dramatic changes in ecosystem processes and functioning. Recent studies have shown that such declines can also affect the transmission of parasitic illnesses. Evidence from Lake Malawi, for example, indicates that overfishing of mollusk-eating fish has resulted in a greater number of *Bulinus* gastropods and the subsequent spread of schistosomiasis (Stauffer *et al.*, 2006).

Schistosomiasis is a parasitic disease afflicting over 200 million people annually. It's carried by freshwater snails. Overfishing may reduce populations of snail predators, resulting in a greater risk of human schistosomiasis.

Biodiversity and Medicines

Decreasing biodiversity is a problem for people, no matter where or how they live. For people living rural subsistence life styles, the extinction of culturally important species means the loss of livelihood and traditions that have supported them for centuries. Loss of biodiversity can lead to loss of health due to malnutrition, and an increased incidence of poor childhood development and nutrition-related diseases. Over the course of history, humans have consumed around 80,000 edible species of plant and animal, and around 3,000 of those have been used widely. We need somewhere between 50 and 100 different vitamins, minerals, and other identified and unidentified substances to stay healthy. The majority of people don't get all these nutrients from their current diets, and reduction in biodiversity and over simplification of food plays a big role (Emma, 2008).

Animals, plants, and microorganisms are a major source of medicines to treat disease. They have already provided us with treatments for such major afflictions as cancer, heart disease, hypertension, inflammatory disorders, and a range of bacterial, fungal, and viral infections. Unani, Ayurveda and many indigenous medicines also utilize animals and their parts or extracts as remedies for various diseases.

An estimated 2000 tonnes of herbs are used annually in India, and the ratio of traditional healers to western-trained doctors reaches 150:1 in some African countries. Many countries, such as Thailand, Sri Lanka, Mexico, China and India, have integrated traditional medicine into their national health care systems. There is also a very large and expanding commercial trade in medicinal plants, involving an estimated 2,500 species. In addition to the direct harvest for traditional medicines, biodiversity also provides both information and raw materials that underpin medicinal and health care systems worldwide in the formal sector. More than half of the world's modern drugs are derived from biological resources. About 85 per cent of traditional medicines involve the use of plant extracts. Although the number of plant species that have been used for medicinal purposes is not known accurately, it is estimated that current use exceeds 50,000 species, including almost 20 per cent of the Chinese flora, around 7,000 species in India, and some 10 per cent of Indonesia's flora. The use of other groups in traditional medicine is infrequent, and poorly documented. Estimates of the number of marine species used for medicinal purposes ranges from a few hundred to a few thousand, the use of which is mainly confined to Asia (UNEP-WCMC, 2007).

Yet only a tiny fraction of biological species has been studied for potential therapeutic effect. Only 2-5 per cent of the estimated 250,000 species of higher plants have been studied – and we are currently losing about one plant species each day. Of the 150 most commonly prescribed drugs in the United States, 57 per cent contain at least one major active compound derived from, or patterned after compounds from nature. In US pharmaceutical market, ten most frequently prescribed drugs are listed in Table 9.1 (Grifo *et al.*).

Effect on Nutrition and Food

Biodiversity plays a crucial role in human nutrition through its influence on world food production, as it ensures the sustainable productivity of soils and provides the genetic resources for all crops, livestock, and marine species harvested for food. Access to a sufficiency of a nutritious variety of food is a fundamental determinant of health. Nutrition and biodiversity are linked at many

Table 9.1: Ten Most Frequently Prescribed Drugs in the United States

Rank	*Drug Name (Generic)*	*Use*	*Surface Organism*
1	Conjugated estrogenes	Estrogen replacement therapy	Animals
2	Amoxicillin	Antibiotic	Fungus
3	Ranitidine	Ulcers	Animals
4	Nifedipine	Heart disease/high blood pressure	Synthetics
5	Levothyroxine	Thyroid hormone replacement therapy	Animals
6	Digoxin	Heart disease	Plants
7	Alprazolam	Antianxiety/sedative	Synthetic
8	Enalapril maleate	High blood pressure	Animals
9	Cefaclor	Antibiotic	Fungus
10	Amoxicillin+ clavulanic acid	Antibiotic	Fungus+bacterium

levels: the ecosystem, with food production as an ecosystem service; the species in the ecosystem and the genetic diversity within species. Nutritional composition between foods and among varieties/cultivars/breeds of the same food can differ dramatically, affecting micronutrient availability in the diet. Healthy local diets, with adequate average levels of nutrients intake, necessitates maintenance of high biodiversity levels.

Conclusion

Effects of disruptions to ecosystems on biodiversity losses are very diverse and remain largely unstudied. It is therefore difficult to quantify current and future health effects of biodiversity losses. Recent studies at the interface of biodiversity and health are helping to elucidate how changes in biological diversity affect health-related outcomes, but policies that are derived from basic research still need to be designed and implemented.

In response to these serious challenges, this chapter adds to our understanding of the complex links between biodiversity and human health, with the hope that this knowledge will lead to decisions that maintain the best possible balance.

References

Allan, B. F., Keesing, F. and Ostfeld, R. (2003). Effects of forest fragmentation on Lyme disease risk. *Conse. Bio.*, 17(1), 267-272.

Behera, K.K., Sahoo, S. and Patra, S. (2008). Floristic and medicinal uses of some plants of chandaka denudated forest patches of Bhubaneswar, Orissa, India. *Ethnobot. Leaflets* 12: 1043-1053.

Brownstein, J. S., Skelly, D. K., Holford, T. R. and Fish, D. (2005). Forest fragmentation predicts local scale heterogeneityof Lyme disease risk. *Oecologia* 146:469–475.

Emma Lloyd. (2008). Is Medical Science Hindered by Loss of Biodiversity? *Laurie Patsalides* http://www.brighthub.com/environment

FAO. (2008). Climate change and biodiversity for food and agriculture. Food and Agriculture organization, Rome.

Grifo, F and Rosenthal, J. Biodiversity and human health. Center for Biodiversity and Conservation American Museum of Natural History.

Marvier, M. (2007). Pharmaceutical crops have a mixed outlook in California. *California Agricul.*, 61(2): 59-66.

Rapport, D. J. and Lee, V. (2003). Ecosystem approaches to human health : Some observation on North/South experiences, *Ecosystem Health*, 3: 26-39.

Rapport, D. J. and Merger, D. (2004). Expanding the practice of ecosystem health. *Ecosystem Health*, 1 (2):4-7.

RCF. Causes of recent declines in biodiversity. Rain Forest Conservation. www.rainforestconservation.org.

Selliers, J. (2005). Facts on Biodiversity. Millennium Ecosystem Assessment.

http://www.greenfacts.org/en/biodiversity/biodiversity-foldout.pdf

Stauffer, J.R., Madsen. H., McKaye, M., Bloch, P., Ferreri, C.P., Likongwe, J. and Makaula, P. (2006). Schistosomiasis in Lake Malawi: Relationship of fish and intermediate host density to prevalence of human infection. *EcoHealth* 3: 22–27.

UNEP. (2010). Integrated Solutions for Biodiversity, Climate Change and Poverty.

UNEP-WCMC (2007). Biodiversity and Poverty Reduction.

USEPA (United State Environment Protection Agency). Pollution. http://www.epa.gov/bioiweb1/aquatic/pollution.html

Vittor, A,Y., Gilman, R.H., Tielsch, J, Glass, G, Shields, T.I.M., Lozano,W.S., Pinedo, V. and Patz, J. A. (2006). The effect of deforestation on the human-biting rate of *Anopheles darlingi*, the primary vector of falciparum malaria in the Peruvian Amazon. *Am. J. Trop. Med. Hyg.* 74: 3-11.

Walsh, J.F, Molyneux, D. and Birley, M.H. (1993). Deforestation: Effects on vector-borne disease. *Parasitology* 106 (suppl.): S55–S75.

WHO. Biodiversity. World Health organization.

Yasuoka, J. and Levins, R. (2007). Impact of deforestation and agricultural development on anopheline ecology and malaria epidemiology. *Am. J. Trop. Med. Hyg.*, 76: 450–460.

Chapter 10

Law Relating to Biodiversity

☆ *Vikram Singh Choudhary &*
Anuradha Choudhary

Environmental law, rules and regulations are arguably a well-developed area of law in India. A new entrant into this field of legislation is the proposed National Biological Diversity Act. Like some of the other environmental laws[1], the impetus and mandate for this law too springs up from an international convention, namely, the Convention on Biological Diversity (CBD). The stated objectives of the CBD are: biodiversity conservation, sustainable use of biological resources, and equitable sharing of benefits arising from such use. One of the key aspects of the CBD is its policy and mandate with regard to Access to Genetic Resources. It has been argued that the agenda behind the CBD and its provisions on access was simply to legitimize access to and control of the genetic resources of gene-rich countries. However, it is also widely recognized that whatever may have been the politics behind the evolution of the CBD, it does have certain strong provisions which ensure that the genetic resources and knowledge associated with the same can no longer be treated as a 'free good', and that there is scope for the framing of regulations for controlling access to such resources in the national and local community interest. At the basis of these developments however is the inescapable realization that genetic resources have tremendous economic potential which is being sought to be harnessed. In the process, like any other economic activity, there would be implications for rights– to property, over use of resources, to knowledge, the possibilities of clashes between cultures, between different systems of medicine and different agricultural practices, and a whole plethora of basic definitional issues regarding: 'ownership', 'access', 'rights to knowledge and resources', 'conservation', 'sustainability', 'equity', 'informed consent' and 'benefits'. How law and policy should respond to this is the challenge.

Rights to the Resource

The CBD recognizes that the State has sovereign rights over its biodiversity. For the purpose of law at the national level, provisions relating to access and benefit sharing in the CBD raise issues relating to property and use rights over land and things that grow on the land. In the first instance, the question that arises is who has rights of ownership over biodiversity and over its constituent genetic resources? Should this be the sovereign right of the state/or the property of the private owner of the land where such resource occurs/or of the community that lives amidst such resources, or of a combination of all of these?

The law relating to property in India recognizes that things attached or rooted to the earth constitute immovable property. It also recognizes that 'benefits arising from land' is an interest in the land and therefore immovable property. But would "benefits arising out of land" include the benefits arising from use of genetic material components in the biological resources found in the land? In cases of land belonging to a private owner, it would seem that biological resources accessed from it are the property of such owner, and can be taken only with the permission of such owner. What would be the law in cases when the land belongs to a person, whereas another is responsible for the conservation and sustainable use of the resource? What would be the law when land belongs to one person, and another has knowledge and information regarding the resources on the land? These aspects are yet to be answered. These questions become particularly more critical when the reality of land ownership in India is examined.

In cases where the land from which access is sought belongs to a village, the relevant law to be consulted would be the Panchayat Acts. Most Panchayat Acts however do not vest 'ownership' over land with the village, although matters such as 'land improvement', 'soil conservation', 'maintenance of community assets', 'development schemes for the village', are matters frequently listed in Panchayat Acts. The Panchayat (Extension to Scheduled Areas) Act, 1996 has stronger provisions that state that the Gram Sabha (village assembly) is empowered to safeguard the traditions and customs of the people, their cultural identity and community resources and further that tribals would have ownership over minor forest produce in such areas.

A substantial part of the biodiversity in India exists in the "Protected Areas" declared under the Wild Life (Protection) Act, 1972, or in the Reserved and Protected Forests under the Indian Forests Act, 1927. The jurisdiction over these areas vests with the State Forest Department. Neither of these laws addresses the issue of accessing the genetic resources within their jurisdiction. In general however, the Indian Forest Act, 1927 states that access to the forest resources can be had with the written permission of the Forest Officer or by any rule made by the State Government. The Wild Life Protection Act, 1972 states that plants may be removed from the area under their jurisdiction only with the permission of the Chief Wildlife Warden under the circumstances specified in the Act. Such circumstances are generally collection of such plants for education and research purposes, or for management purposes such as fire control. Access to and collection of the resources for commercial purposes is therefore not contemplated under the Wild Life Protection Act, 1972.

Both the Indian Forest Act, 1927 and the Wild Life Protection Act, 1972 also provide for the settlement and recognition of the rights of the people dependent on the resources in areas within their jurisdiction. Where such rights are recognised, should such people have a say in access to such resources by 'outside' interests? Where such rights of communities to collect and use a resource were traditionally and customarily recognised but are not so by existing provisions of law, what would be the status of such rights? Who should decide on the access when the knowledge of the users of the biological resource in a forest land is the basis on which physical access to the resource is sought? The problem with regard to access to the resource in all the scenarios contemplated above is complex because access to the physical resource is more often than not, not sought at random. The reason and basis for the access in the first place is the existing knowledge and information regarding the resource. In the absence of any legal forms to protect that knowledge, the question that really has to be addressed are on what basis should access be regulated?

The proposed law relating to biodiversity drafted by the Ministry of Environment and Forests works on CBD's premise that the State has the sovereign right over its genetic resources. The law proposes to establish authorities at the national, state and local levels to deal with the issues of access to genetic resources. However, the issues with regard to 'ownership', jurisdiction, and inter-play with existing laws are yet to be addressed and resolved. For instance, can the National Authority permit access to resources within the jurisdiction of a national park or sanctuary, and can it permit commercialisation of the same?

The Draft Law on Biodiversity

A draft law on Biological Diversity has been drafted. In essence the law puts in place a mechanism for regulating access to genetic resources. It addresses several of the issues raised above. Some of its key features are:

- The Act states as its objectives the conservation of biological diversity, sustainable use of its components, and equitable sharing of benefits arising out of the use of biological resources. The term biological resources are defined as–plants, animals and micro-organisms and parts thereof, and their genetic material and by-products, with actual or potential use or value, but does not include human genetic material.
- The Act proposes to set up bodies at three levels, to carry out its functions. At the national level, there will be a National Biodiversity Authority (NBA), which will screen proposals for transfer of genetic resources abroad, advise the central government of measures for conservation,

sustainable use, and benefit-sharing, suggest the use of the National Biodiversity Fund, and oppose, where necessary, IPRs in India and abroad which violate the Act's provisions. A proposed provision seeks to ensure that where patents are sought to be granted on any 'invention' based on research and information on a biological resource occurring in India, the Controller of Patents shall refer it to the NBA for its permission and any other conditions that it may stipulate.

- At the state level, there will be State Biodiversity Boards (SBB), which will oversee use and conservation of biodiversity within state jurisdiction. SBBs will also manage the State Biodiversity Funds. At local levels, there will be Biodiversity Management Committees, which will have a voice in regulating the transfer, use, and conservation of resources and knowledge at community and individual level. There is also provision for creation local biodiversity funds at the level of institutions of local self-government. Rules are to be framed in respect of all the funds under the proposed law.
- As regards access to genetic resources, the law prohibits any person who is not a citizen of India, any body corporate, association or organization which is not registered in India, or which is registered in India but has non-Indian citizen participation in equity or management, from obtaining any biological resource occurring in India and/or associated knowledge for research, commercial utilization, or bio-survey or bio-utilization without the prior approval of the National Authority. Approval of the National Authority is also required in the context of transfer of material from any citizen of India to a non-citizen or non-resident or body corporate having non-Indian participation.

  With regard to access by Indian citizens/body corporate registered in India with Indian participation alone, the standards for approval are much lesser, and such entities are only required to give prior intimation to the concerned state Biodiversity Board before utilization for commercial or research purposes. Approval from either the State or the National Authority is, however, not mandated. The State Biodiversity Board *may* restrict the activity of the same is found violative of the objectives of conservation, sustainable use and benefit sharing.

The difference in treatment between Indian and non-Indian entities raises questions about the basis of this distinction, since Indians are not necessarily going to behave more responsibly towards the resource, or be more respectful of indigenous knowledge, when given easier access.

- ☆ While there is provision that Biodiversity Management Committees at the local level would have to be consulted by the National Authority and the State/Union Territory Biodiversity Boards in decisions relating to the use of biological resources and associated knowledge within their jurisdiction, there is no clear provision as regards how access to and use of information and knowledge of an identifiable community would be affectuated. For instance, if access is sought from a community whose representation at the local body is in minority, there is the possibility that its voice might get over-ridden. Further, there is no requirement for obtaining the informed consent of the affected community/individual whose knowledge and information is being accessed. The law further provides that 'where necessary', the National Authority shall give information regarding approvals through public notice; however, there is no provision for 'public hearing' and addressing grievances through such hearings. The only remedy available would be through petitioning a court of relevant jurisdiction, after a notice of 60 days has been given to the Central Government in this regard. The State Biodiversity Board is not required to give notice of any kind for applications considered by it.
- ☆ The law mandates the National Biodiversity authority to challenge IPRs within India and abroad, which are found to be violative of India's and its communities' rights.

The provisions of the biodiversity law provide the basic framework for achieving control over access to genetic resources in the interests of achieving conservation, sustainable use and assuring benefit sharing from use of the resources. However, there are still certain aspects that need to be addressed, and it is hoped that this would be done in the Rules to be framed under the law. The basic problem, evident as it is, is that rules do not have the sanctity of the provisions of the statute. Rule making is an administrative act which does not need to confirm to the rigorous standards of debate and discussion as a law. By that same logic, rules are far easier creatures to amend.

Some of the tasks for the Rules would be to achieve the following:

1. Terms and Conditions for Access

The terms and conditions under which approval for access would be given for 'research purposes', and for purposes of 'commercialization'. Monitoring and reporting requirements mandating the person seeking access to keep the community/ country of origin informed of the developments from the research on the material.

2. Ensuring Conservation

The manner in which such approval would be given by the relevant authorities, and the terms and conditions that would be prescribed by them, keeping in view specific aspects such as the nature of the resource being accessed- an assessment of the ecological implications of its access, and pointers towards how its conservation and sustainable use can be achieved.

3. Achieving Prior Informed Consent

In cases where access to resource and information is sought from an identifiable community/group of persons within a community/individual within a community (commonly referred to as the 'originator'), the rules would have to be more elaborate in terms of addressing concerns such as: (*a*) the "informed consent" of the community–how such consent may be achieved–for instance prescription of requirements such as communication of the bio-prospecting activity in detail in the local vernacular language of the originator, interactive sessions and public hearings through which the originator understands the nature and implications of the access sought; (*b*) what would be the terms for continued involvement of the originator in the bio-prospecting activity and for achieving conservation of the resource; (*c*) what would be the terms of benefit sharing, and how would the involvement of the originator in such negotiation be achieved. Model Material and Information Transfer

Agreements would also have to be drafted to incorporate the basic elements for achieving the above.

4. Achieving Benefit Sharing

What kind of benefits are appropriate when a commercial product is derived from traditional knowledge: a 50:50 share of profits between the company and the originator which held the knowledge, the transfer of relevant technologies to this community, participation in R&D, exemption from the application of IPRs, revival of tenurial rights to land and resources, non-monetary awards, or others? By what mechanism would beneficiaries be identified, and benefits be transferred? What if more than one community holds the same knowledge, should benefits be shared with all? What in the case of innovations made by single individuals, or families, within a community: should they get the benefits, or the whole community?

Applicability of General Principles of Law

Constitutional Law

As mentioned earlier, one of the interesting aspects of the practice of the law is the testing of any statute, rule or administrative action against the basic tenets of the Constitution. The effort in this section shall be to highlight some of the constitutional principles that can be brought to come into play in the context of: (a) the rules and regulations for access, and administrative action of the authorities constituted under the proposed Biological Diversity Act; and (b) the legal principles applicable to any private party seeking access.

The chapter on Fundamental Rights[2] and Directive Principles[3] in the Constitution elucidate the basic tenets of individual liberty, the reasonable restrictions on the same in view of public interest, and the principles that shall constitute the basis for governance by the State. Some of the fundamental rights of relevance to the topic under discussion are:

- Equality before the law which has been interpreted as the right against unreasonableness and arbitrariness[4]
- Right to life and personal liberty which has been interpreted as the right to livelihood, life with human dignity[5]
- Right to conserve distinct language, script or culture[6]

- Right to enforcement of fundamental rights through writ proceedings at the Supreme Court[7] or any High Court[8]

Constitution further provides that it shall be duty of the State to apply the Directive Principles in the making of laws. Some of these principles of interest to the present debate are:

- The State shall strive to promote the welfare of the people by securing and protecting as effectively as it may a social order, in which justice, social, economic and political shall inform all institutions of political life[9].
- The State shall, in particular, direct its policy towards securing that the ownership and control of the material resources of the community are so distributed as best to sub-serve the common good[10].
- The State shall promote with special care the educational and economic interests of the weaker sections of the people, and in particular of the Scheduled Castes and Scheduled Tribes and shall protect them from social injustice and all forms of exploitation[11].
- The State shall ensure that opportunities for securing justice are not denied to any citizen by reason of economic or other disabilities[12].

Some of the principles laid down by the judiciary in interpreting the above are as follows:

- Our Constitution makes it imperative for the State to secure to all its citizens the rights guaranteed by the Constitution and where the citizens are not in a position to assert their rights. The State must come into the picture and fight for the rights of the citizens. The Preamble to the Constitution, read with the Directive Principles, Articles 38, 39 and 39A enjoin the State to take up these responsibilities[13].
- Economic empowerment is a basic human right and a fundamental right as part of right to live, equality and of status and dignity to the poor, weaker sections, dalits and tribes. Justice is an attribute of human conduct and the rule of law is an indispensable foundation to establish socio-economic justice[14].

- Duty to act fairly is part of fair procedure envisaged under Articles 14 and 21. Every activity of a public authority *or* those under public duty or obligation must be informed by reason and guided by the public interest[15].
- Another interesting aspect is the willingness of the Judiciary to impose strictures on a private entity if the activities of such entity infringe upon the fundamental rights of the people. It may therefore be a possible argument that the activity of a person seeking access to genetic resources, by its very nature has implications for the rights of people. The obligations on such a person, for instance to follow all the elements of the PIC procedure, achieve mutual benefit sharing, etc., should therefore be enforceable not just on the basis of a contractual obligation or a statutory obligation, but as a 'public duty', enforceable by a writ of mandamus. It has been held by the Supreme Court that a writ of mandamus may lie against a private entity depending on the nature of duty imposed on that entity. The duty must be judged in the light of the positive obligation owed by the person or authority to the affected party, no matter by what means the duty is imposed.[16]
- In the absence of suitable laws, international conventions and norms, so far as they are consistent with the constitutional principles, can be relied upon by the judiciary[17].

In view of the directive principles for securing social welfare and common good, absolute freedom of contract and of laissez faire are no longer valid principles[18].

Contract Law

A study of some of the principles of contract law reveals how these could be used to address some of the concerns regarding unequal bargaining positions addressed in this chapter. For instance, the definition and interpretation of "free consent" under the Indian Contract Act, 1872, could provide valuable pointers to the development of PIC under the legal regime for biodiversity:

- Section 14, Indian Contract Act, 1872 defines "free consent" as: "Consent is said to be free when it is not caused by coercion, undue influence, fraud misrepresentation, mistake."[19]

- A contract is said to be induced by "undue influence" where the relations subsisting between the parties are such that one of the parties is in a position to dominate the will of the other and uses that position to obtain an unfair advantage over the other.[20]
- Where a contract appears on the face of it or on evidence adduced to be unconscionable, the burden of proving that such contract was not induced by undue influence shall lie upon the person in a position to dominate the will of the other[21].
- Fraud is defined as acts committed by a party to a contract with intent to deceive another party to enter into such contract.[22]
- The term 'misrepresentation' is defined as including causing, however innocently, a party to an agreement to make a mistake as to the substance of the thing which is the subject of the agreement[23].

Conclusion

The above issues are a small part of the larger range of questions that arise in the 'access debate'. The so-called 'promises' of genetic engineering and growth of the biotechnology industry are one of the basis of the access debate. And while the promises are many, the problems are equally bothersome. Biotechnology, as with any other technology, is a major source of public power in modern society, and raises important concerns regarding rights of use, control and participation in its use.

The recent controversy surrounding the field testing of genetically modified cotton, and debates on implications of technologies popularly called the 'terminator' and 'traitor' technologies raise further challenges for our laws relating to the nature of standards to be adopted for making an 'informed' choice about the environmental and health implications of such technologies and use of its products. The growing use of genetically modified crops raises more issues- about the safety of transferring organisms into new environments, questions of liability for damage, and the need for greater transparency in obtaining information. Labeling of genetically modified products to enable the consumer to have a choice, is another issue for the law. The applicability of the precautionary principle to the assessment of such technology does not have endorsement of all countries as yet.

Some questions posed by the Human Development Report, 1999, could be effective guides to the choices that law and policy would have to make in this whole debate:

"Does the control, direction and use of technology:

- ☆ Promote innovation and haring of knowledge?
- ☆ Restore social balance or concentrate power in the hands of a few ?
- ☆ Favour profits or precaution?
- ☆ Bring benefits for the many or profits for the few?
- ☆ Respect diverse systems of property ownership?
- ☆ Empower or disempowering people?
- ☆ Make technology accessible to those who need it?"

References

[1] The Air (Prevention and Control of Pollution) Act, 1981 and the Environment (Protection) Act, 1986 were enacted under Article 253 of the Constitution of India, 1950 as legislations to give effect to a decision adhered to at an international conference, namely the United Nations Conference on Human Environment held in Stockholm in June 1972.

[2] Part III, Constitution of India, 1950.

[3] Part IV, Constitution of India, 1950.

[4] Article 14, Constitution of India, 1950.

[5] Article 21, Constitution of India, 1950.

[6] Article 29, Constitution of India, 1950.

[7] Article 32, Constitution of India, 1950.

[8] Article 226, Constitution of India, 1950.

[9] Article 38, Constitution of India, 1950.

[10] Article 39(b), Constitution of India, 1950.

[11] Article 46, Constitution of India, 1950.

[12] Article 39A, Constitution of India, 1950.

[13] Charan Lal Sahu v. Union of India (1990) 1 SCC 613.

[14] Muralidhar Dayandeo Kesekar v. Vishwanath Pandu Barde 1995 Supp (2) SCC 549.

[15] LIC of India v. Consumer Education and Research Centre (1995) 5 SCC 482.

[16] Andi Mukta Sadguru Shree Muktajee Vandas Swami Suvarna Jayanti Mahotsav Smarak Trust v. V.R.Rudani (1989) 2 SCC 691. Also see, K.Krishnamacharyalu v. Sri Venkatashewara Hindu College of Engineering (1997) 3 SCC 571.

[17] Visakha v. State of Rajasthan (1997) 6 SCC 241.

[18] Y.A.Mamarde v. Authority under the Minimum Wages Act (1972) 2 SCC 108.

[19] Section 14, Indian Contract Act, 1872.

[20] Section 16(1), Indian Contract Act, 1872. Taking unfair advantage of one's economic power has been held to be unconscionable under English law: Lloyds Bank Ltd. v. Bundy (1974) 3 All E.R. 757; Schroeder Music Publishing Co. v. Macaulay (1974) 3 All E.R. 616.

[21] Section 16(3), Indian Contract Act, 1872.

[22] Section 17 Indian Contract Act, 1872. 'Fraud' includes acts of: suggestion as a fact of that which is not true by one who does not believe it to be true; the active concealment of a fact; a promise made without intent to perform it; any act fitted to deceive.

[23] Section 18, Indian Contract Act, 1872.

Chapter 11

The Impact of Biotechnology on Agrobiodiversity Conservation

☆ *Arpita Awasthi, Rashmi Arnold & Deepak Mishra*

Agrobiodiversity refers to biodiversity related to agriculture and can be described as 'the variety and variability amongst living organisms that are important to food and agriculture in the broad sense and associated with cultivating crops and rearing animals and the ecological complexes of which they form a part. It includes the diversity found in farming systems as well as their surroundings to the extent that the latter influences agriculture. The relationship between biotechnology and agrobiodiversity also largely holds for these technologies. The interactions between biotechnology and agrobiodiversity are manifold. Traditional biotechnological applications, in particular, make use of microbial organisms. They include:

- ☆ Composting, the accelerated microbial degradation of organic matter.
- ☆ Nitrogen fixation, based on the ability of bacterial symbionts of leguminous plants to fix atmospheric nitrogen.
- ☆ Fermentation, generating alcoholic beverages and dairy products and preserving a large variety of meat and plant products.
- ☆ Ethnoveterinary practices aimed at protection of domestic animals against infectious diseases.

For these applications, specific microbial strains have co-evolved and often been selected. In other words, biodiversity has been exploited to allow for these traditional biotechnological applications. Modern biotechnology has developed much more refined tools to select and generate optimal microbial organisms for a larger set of applications, and each of these involves the use of specific traits generated and expressed in the microbial domain.

These applications, as well as applications making use of plant and animal biodiversity, have contributed to the shaping of agricultural production. Cereals and pulses have been intercropped in all Centers of Origin of agricultural crops, an essential combination to guarantee high yields. Other crops, such as soybean and enset (Ethiopia), have gained importance because protocols were developed to ferment the harvested products and allow for prolonged food storage. Ruminants (cattle as well as sheep and goats) have been selected for milk production concomitantly with the development of milk fermentation technologies. Modern biotechnology has already influenced the genetic diversity of crops and animals cultivated and raised in the fields by allowing for the rapid and wide-spread introduction of desired starting material through *in vitro* technology. It is now on the brink of more profoundly changing our agriculture, in particular by the generation of novel crop varieties containing traits which could not be incorporated before and altering our farming practices. In this way, biotechnology has also changed and will further change our agricultural biodiversity. It can improve our production systems and the diversity of products we desire. Modern biotechnology also has a potential or actual influence on natural ecosystems, either by allowing changes in agriculture which affect natural biodiversity or by directly influencing natural biodiversity itself.

From ancient times onwards, farming practices, including animal and plant breeding, have profoundly influenced our agro-ecosystems and the biodiversity these contain, and thus this influence is not the prerogative of biotechnology but of technologies applied in agriculture in general. Impacts on biodiversity may not be different in nature, only in degree, from those of our traditional or conventional practices. The term genetic resources refer to a resource-centered view on biodiversity. Accordingly, it is often stated that loss of biodiversity means loss of capital and potentially useful resources. The immediate relevance of agrobiodiversity concerns food production and the prime incentive to preserve agrobiodiversity is based on economics.

Biotechnology has provided tools to more effectively maintain and utilize genetic resources *ex situ* as well as on-farm, and has thus contributed positively to the conservation of genetic resources. A limited contribution of biotechnology to the development of novel genetic resources in agro-ecosystems may also emerge in the near future, as a result of introgression of novel traits by means of genetic modification. However, the increasing role of biotechnology also forms a potential threat to the survival of agrobiodiversity in the environment, whether this is in the farmer's field (on-farm), or in natural ecosystems (*in situ*). Biotechnology can be regarded as the latest factor contributing to this industrialization and changing our food production and, in parallel, our agrobiodiversity.

In vitro technology allows for the rapid propagation of plant material. In tissue culture, plant parts or plantlets are grown under sterile conditions and split in several subsamples at regular intervals. This material can then be regenerated and reintroduced to *in vivo* conditions, resulting in a fast multiplication of the plant material. Such tissue culture approaches are in particular and successfully used for root and tuber crops, such as potato, sweet potato, cassava, as well as banana. Artificial insemination (AI) of domestic animals is generally used to promote the spread of genetic properties of a very small number of animals. Sperm of a recent Dutch top bull belonging to the dominant Friesian-Holstein race has been used in over a million inseminations worldwide. *In vitro* fertilisation (IVF) is another capital intensive technology in animal breeding only used for a narrow, highly valued set of breeding materials. This utilization of the technology has helped in acknowledging the value of farmers' varieties, in other words of on-farm managed biodiversity.

Biotechnology contributes to a growing uniformity in food production; it will also result in a growing uniformity in food consumption. Novel varieties in staple crops improved with the use of molecular markers or containing transgenes will outcompete and largely replace traditional varieties because of a greater abundance and lower prices. Traditional varieties will only survive where these have distinct added value, in particular in processing and taste or in their use in cultural practices. Molecular markers may help protect traditional varieties and regional products as these act as an instrument to monitor and control whether a product which is marketed for its specific quality does indeed adhere to the agreed standards or not, and to prevent misappropriation of traditional or modern farmers' varieties by third parties.

References

Almekinders C and De Boef W (2000). Encouraging diversity. The conservation and development of plant genetic resources. Intermediate Technology Publications, London.

Bunders J, Haverkort B, Hiemstra W (1996). Biotechnology; Building on Farmers' Knowledge. Macmillan Education Ltd, London, UK.

FAO (1996). Report on the State of the World's Plant Genetic Resources for Food and Agriculture. FAO, Rome.

Human Development Report 2001. United Nations Development Programme: Sustainable Development. Press Kit.

Lipton M (1999). Reviving the stalled momentum of global poverty reduction: what role for genetically modified plants. Sir John Crawford Memorial Lecture, CGIAR International Centres Week, CGIAR Secretariat.

Persley G J, Lantin MM (1999). *Agricultural Biotechnology and the Poor.* National Academy of Sciences Consultative Group on International Agricultural Research, Proceedings of an International Conference, Washington D.C.

Persley GJ, Lantin MM (2000). Agricultural biotechnology and the poor: proceedings of an international conference, Washington, D.C., CGIAR, Washington, USA..

Section II
Research Papers

Chapter 12

Conservation of Biodiversity in the Natural Forests of Central India: A Case of Few Critically Endangered Medicinal Species of Mandla and Bhopal Region

☆ *Manish Mishra*

The unscrupulous collection of medicinal plants from wild habitats by various stakeholders not only threatened medicinal plant resources but also deteriorated quality. An adequate quality check has become a dire necessity by which the consumer at large may be assured that the products they use do not contain toxic ingredients or have different therapeutic action. The increase in demand of medicinal plants for the commercial herbal medicine sector led to indiscriminate and unscientific collection without any consideration for the quality of the material collected. In many cases the immature extraction of fruits, roots, tubers etc. has drastically reduced the quality as well as quantity of the raw product to the below critical level. Many studies on different species have confirmed the same. Reports indicate that due to immature extraction, population, per plant yield and quality of raw material is declining in the natural forest of central India (Prasad *et al.*, 2002, 2003 and Mishra *et al.* 2004, 2005).

Ecological status of selected 40 NTFP in Madhya Pradesh was assessed through a CAMP (Conservation Assessment and Management Plan) workshop in June, 1998 at Indian Institute of Forest Management, Bhopal. This assessment was based on IUCN categories of 1994. This was done by involving field foresters, academicians and other stakeholders. Out of 40 assessed species from the state of Madhya Pradesh which accounts for 60 per cent of total recorded MAPs wealth of the country, 2 species (*Curcuma caesia, Rauwolfia serpentina*) were found in the category of critically endangered; 9 species were found to be endangered; 14 species vulnerable; 9 species with lower risk least concern and one (*Jatropha curcus*) could not be evaluated (Prasad and Patnaik, 1998).

The IIFM has conducted the detailed aspects of sustainable harvesting, value addition, processing and marketing of some important medicinal plant species, like Aola (*Emlica offinalis*), Bel (*Aegle marmelos*), Satawar (*Asparagus rasimosus*), Kullu (*Sterculia urens*) in the tropical dry deciduous Forests of Sheopur Forest Division (Bhattacharya *et al.*, 2002) to understand the socioeconomic factors linked with unsustainable removal of these product. A study on *Emblica officinalis, Buchanania lanzan, Chlorophytum* spp. and *Terminalia arjuna* (bark) conducted in the Central part of Madhya Pradesh revealed that due to destructive premature harvesting the regeneration of these species has been adversely impacted and productivity is declining (Prasad *et al.*, 2000). These unsustainable practices are adversely impacting the biodiversity. Prasad *et al.* (2002), Mishra, 2007 studied the harvesting practice of safed musli (*Chlorophytum* spp.) and reported that due to unsustainable extraction the population of this species is declining in the natural forests as well as the protected areas of Panna and Satna districts (M.P.). The natural regeneration of this species is also affected due to immature harvesting. Mishra (2001) has also reported that present harvesting practices of *Curcuma caesia* and *Rauwolfia serpentina* in the natural forest of M.P are unsustainable. It was further reported that both the species occurring in the forests are gradually getting depleted due to over harvesting and consequently poor regeneration.

Mishra *et al.*, (2004) reported low density and poor yield of Kali musli and Safed musli in Bhopal and Mandla forest divisions owing to unsustainable collection practices. They reported that natural population of safed musli is decreasing at alarming rate while Kali musli is relatively common due to less market demand and consequently lesser extent of extraction. Lot of work has been done by workers at IIFM, Bhopal on the wild harvesting of NTFPs including herbaceous medicinal plants of central India and its impact on density, regeneration and harvesting under the natural forest conditions. They reported a sharp decline in the per hectare density, occurrence of seedling and yield of roots/fruits/leaves etc. due to immature collection, competition among villagers to collect as much as possible, market and traders pressure and increasing demand by domestic, Ayurvedic, pharmaceutical industries. (Kotwal & Mishra 2007, Mishra & Teki 2007, Mishra & Kotwal 2007, Mishra, 2009, Mishra & Kotwal 2009).

Study Site

Bhopal District

Bhopal district is situated on 23°16′ North latitude and 77°25′ East longitude. The total area of forest is 437.15 sq. km which, about 15.77 per cent of total geographical area of 2778.0 sq. km. The altitude varies from 450 to 600 m above mean sea level. The topography of the Malwa plateau presents undulating surface interspersed with areas of rich black cotton soil. The climate of Bhopal is monsoonic indicating a seasonal rhythm of weather during the year, which may be divided into four major seasons. The period from March to the middle of June is the summer season. October and November constitute the post monsoon season. The average annual rainfall is reported to be about 919 mm. The forest division is divided into two forest ranges *i.e.* Berasia and Samardha.

Mandla (East Mandla) District

Mandla district is located in the east-central part of the Madhya Pradesh. It lies between the latitude 22° 2' and 23° 22' North and longitude 80° 18' and 81° 50' East. The head quarter of east Mandla forest division is located in the Mandla town. All the forest area fall under division comes under Bicchiya tehsil which constitute of Bicchiya, Mohgaon, Mawai and Gughari development blocks. Mandla is situated 96 km away from Jabalpur by road on Jabalpur – Raipur national highway. According to Champion and Seth revised classification (1968) the forest of East Mandla Forest Division can be broadly classified as Sal, Teak and Mix forest. Mandla district consists of a rugged high tableland in the eastern part of the Satpura hills. The cold season from December to February is followed by the hot season from March to the middle of June. The period from mid-June to September is the southwest monsoon season. October and November constitute the post monsoon or retreating monsoon season. May is the hottest month with the mean daily maximum temperature at 41.3°C and the mean daily minimum at 24°C. On individual days during the summer season the day temperature may go above 44°C.

Research Methodology

The study incorporates both primary and secondary data. The primary data have been collected through various field surveys and the secondary data were sourced from forest department records, village Panchayat record, local Ayurvedic practitioners, IIFM library, Internet etc. Secondary data were also collected from other different sources like published and unpublished literatures, census reports etc. A total of twenty five villages were selected from 2 selected ranges of Bhopal district. The field investigations and survey were conducted during January to December 2009, with the assistance from local people and forest department officials. Random sampling was employed for collection of data and 15 per cent households were surveyed from each selected village.

Density and Regeneration

The five sample plots (transects) in each forest compartment (selected forest ranges) where species was growing were laid randomly. The size of sample plots (transect) was 50 × 5 meters.

The individuals of species in transect was recorded. Smaller size quadrates of 2m × 2m size were laid within each of the transect. All the young seedlings up to a height of 50 cm. were counted in each quadrate.

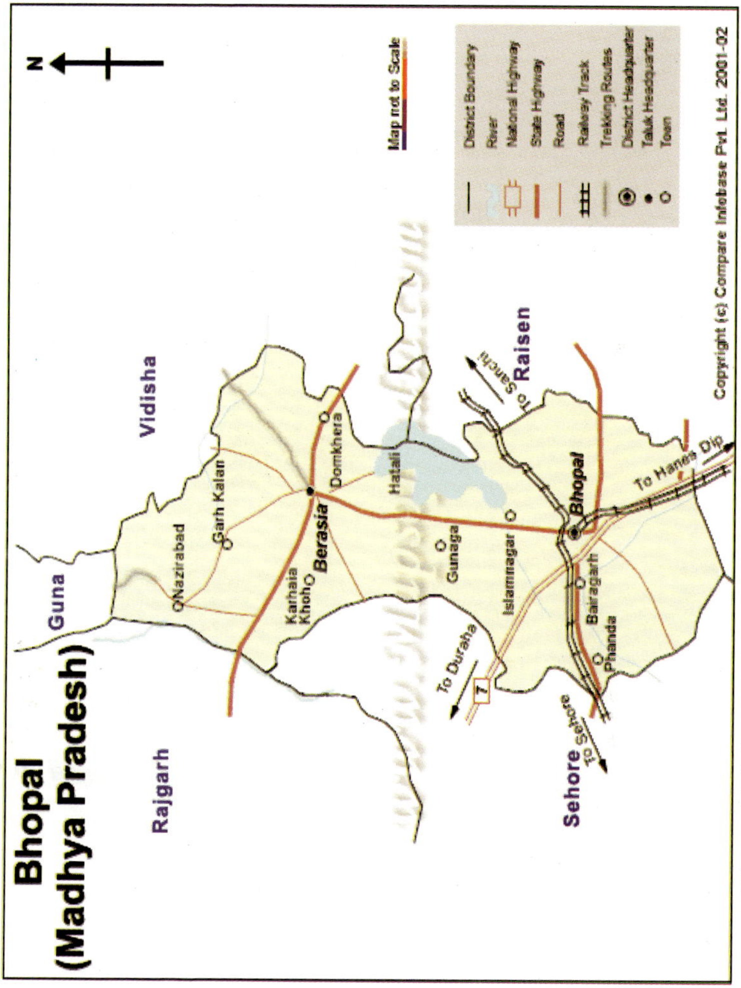

Figure 12.1: Location Map of the Study Area–Bhopal, Madhya Pradesh

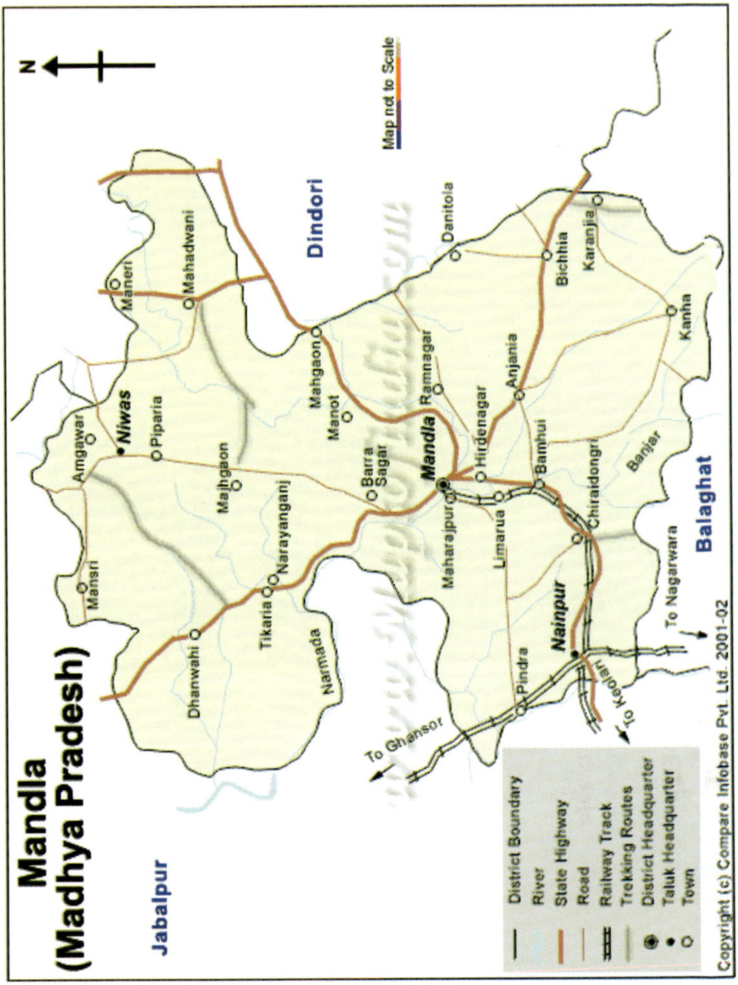

Figure 12.2: Location Map of the Study Area–Mandla, Madhya Pradesh

Method of Plant/Parts Harvesting and Time

Regular field survey and visits to the selected ranges in both the division were conducted at the time of fruit maturation and harvesting. Plant/leaves harvesting period, collection month and mode etc. were recorded minutely during field visits along with the villagers. Present harvesting practices adopted by the local collectors were also recorded. Informations about the harvesting tool, mode etc. were noted during extensive survey of both the divisions. The condition of plants in the natural forests, damage etc. to the plant was also observed during fruit/seed harvesting period.

Questionnaire Survey of Villagers (Gatherers)

A semi structured questionnaire was developed and field tested. From each selected village 15-20 per cent of households were surveyed, based on random sampling technique to know the harvesting method, collection of fruits per day by each person in a season etc.

Causes for Population Decline

Various causes of plant population decline were assessed by collecting data from different sources like forest department personals, local people, NTFP collectors, herbal practitioners etc. regarding previous or earlier years (last decade) conditions of the species in the natural forests. Various field observations on grazing, forest fire incidences etc. were also recorded time-to-time during the field survey.

Results

The data depicted in Table 12.1 shows ecological status of Hadjudi in the study area. It is clear from the data that flowering and fruiting generally occurs in the month of September to December in all the districts with minor difference of one month. It was also noted that stems of plant were collected by the locals only after demand from Ayurvedic industries and big traders of the locality. However, collectors harvest this species based on their domestic demand or for curing their domestic cattle's. Plants were collected immaturely from the natural forest as per the local as well as market demand at anytime in the whole year/season. However, tribal collectors of Mandla are aware of immature harvesting and always take care before collecting Hadjudi from forests.

Table 12.1: Ecological Observations of Hadjudi (*Cissus quadrangularis*) and Present Harvesting Practices in the Natural Forests of Bhopal

Plant Habit	*Place of Occurrences*	**Flowering Month in Natural Forest*	**Fruiting Month in Natural Forest*	**Plant Harvesting Time in N.F. (Maturity)*	*Present Harvesting Time*	*Present Fruit Harvesting Method*
			Bhopal			
Perennial climber	Samardha, Kerwa, Amla, Kolar, Bhanpura	Sept.–Oct.	Oct.–Nov.	Dec.–Jan.	Whole year or as per demand	Uprooting whole plant
			Mandla (East)			
Perennial climber		Sept.–Oct.	Oct.–Nov.	Jan.	Based on demand	Selective uprooting

* Ref: PC Kanjilal (1933). A forest flora for Pilibhit, Oudh, Gorakhpur and Bundelkhand.

The plant density per hectare was observed very less (Avg.>1.00/ha.) while more (Avg.3/ha.) in forest of Mandla (East). On the other contrary, the regenerating individuals were found less than 1/ha. in the natural forests of Bhopal. Poor regeneration was also observed in the natural forests of Mandla East (Table 12.2).

Table 12.2: Plant Density and Regeneration Status of *Cissus quadrangularis* in Bhopal and Mandla District

Name of District	*Plant Density*	*Plant Regeneration*
Bhopal	>1.00/ha.	>1.00/ha
Mandla (East)	3.00	1.00

The ecological status of *Tylophora indica* in the natural forest of study area is shown in Table 12.3. It is clear from the data that flowering and fruiting generally occurs in the month of August-September. It was also noted that leaves were collected by the locals only after demand from Ayurvedic industries and big traders of the locality. However, collectors harvest this species based on their domestic demand. Leaves were collected indiscriminately from the natural forest as per the local/market demand, anytime in the year in the Bhopal district. However, due to sparse distribution of plants, no leaves were harvested from the Mandla.

The density of plant per hectare was observed very less (Avg.>1.00/ha.) in the natural forest of Bhopal whereas significantly more (Avg.4/ha.) in the forest of Mandla (East). On the other hand, the regenerating individuals were absent in the natural forest of Bhopal while 2/ha. were observed in East Mandla (Table 12.4).

Reasons for Population Decline

Unripe Collection

To meet up the demand, the local tribal people resort to over harvest *Tylophora indica* and *Cissus quadrangularis* species from the natural forests. People harvest whole plant in the natural forests of study area during the month of Aug- Sept. while the maturation of rhizome takes place in the months of November-December. Not a single plant was left behind during harvesting of this plant in the natural forests of study area. After complete harvesting, they used to go very far from their villages in search of plant and some time other

Table 12.3: Ecological Observations of *Tylophora indica* and Present Harvesting Practices in the Natural Forests of Bhopal

Plant Habit	*Place of Occurrences*	*Flowering Month in Natural Forest*	*Fruiting Month in Natural Forest*	*Plant Harvesting Time in N.F. (Maturity)*	*Present Harvesting Time*	*Present Fruit Harvesting Method*
			Bhopal District			
Perennial climber	Samardha, Kerwa, Amla, Kolar, Bhanpura	August*	Sept.–Dec.*	Dec.–Jan.*	Whole year or as per demand	Uprooting whole plant, leaves
			Mandla District (East)			
Perennial climber	Sparsely Scattered in all park area	August	Sept.–Dec.	Jan.–Feb.	As per traders demand	Selective plucking of only leaves

* Ref: PC Kanjilal (1933). A forest flora for Pilibhit, Oudh, Gorakhpur and Bundelkhand.

districts. The people have hardly any alternative means to earn livelihood from other sources. The local tribal peoples mostly depend on the forest products including medicinal plants. It is one of the major reasons for population decline of selected species in the studied districts.

Table 12.4: Plant Density and Regeneration Status of *Tylophora indica* in Bhopal District

Name of District	*Plant Density*	*Plant Regeneration*
Bhopal	>1.00/ha.	Nil
Mandla (East)	4.00	2.00

Poor Density

Very low plant density was observed in the unprotected natural forests of the study area. The species mostly found in the patches and does not spread much in the vicinity. Due to over harvesting of the species density is very poor in all the sites and sometimes nil in the whole forest, where earlier reported to be most abundant.

Lack of Regeneration

Almost negligible regeneration was observed in all the districts because of very low plant density of both the plants in the natural forests. Gatherers generally uprooted immature plants before flowering, fruiting; hence regeneration was affected severely in the natural forests of the study area. No seed or any vegetative part of the plant was found on the forest floor of all the selected districts.

Habitat Damage

Habitat destruction is a major external factor affecting plant population under wild conditions. Loss of habitat is also a threat to *Cissus quadrangularis* and *Tylophora indica* growing in the natural forest of all the selected districts. Catastrophes' and human activities to a longer extent are frequently responsible for endangering species through ways like–degradation and fragmentation of habitats, overexploitation of natural resources and catastrophe due to introduced pests and diseases. Smaller populations of *Tylophora indica* and other herbaceous medicinal plants are more prone to extinction and attract less pollinator, less reproductive success and less gene flow.

Cattle Grazing

Uncontrolled cattle grazing are highly detrimental to the selected species. Trampling and grazing by livestock kills most of the young seedlings. This is another decimating factor associated with this species in the natural forests of all the selected districts of Madhya Pradesh. It was found that huge population of livestock depends for grazing on the forest particularly near villages. Although, the domestic cattle are dependent on the forests for grazing but there is more pressure in Bhopal district.

Forest Fires

The frequent forest fire in the study site affect most of the natural forest area of all the forest divisions. These occur more during summer than winter season. The villagers set fire to clear the forest floor, which is littered with combustible dry leaves and twigs. The plant is tuberous and its fruits, seeds dispersed on ground were get burnt due to these forest fires. Forest fires are frequent in all the selected sites.

Some Important Field Observations

It was observed that in the selected district, the plant was found in the fencing of backyard of villager houses and on agricultural bunds, small bushes etc., but not in the dense forests adjoining to the villages. It was informed by the locals that Hadjod is used for curing various animal diseases also. Hence, most of the local people living around the forest grow this plant in their fencing and home gardens. It was also observed that plant propagated through its stem/ vegetative part and their will be no regeneration problem. The plant is hardy and grows on variety of soil and habitat conditions. A little protection is needed for its multiplication, hence, villagers used to grow this plant in their vicinity. On the other hand, *Tylophora indica* grows in specific habitat conditions and requires support of shrubs and trees for its vegetative growth. Generally, species grow in well drained soil with thick canopy cover. Due to its pungent smell (in the leaves), domestic cattle do not graze this plant. However, due to increasing market pressure, the local traders and village middlemen forced harvesters to collect all the leaves from the plant. It was also noted that the leaves were harvested from forest only on the basis of Ayurvedic industry demand. The Ayurvedic manufacturers based on commercial requirement purchased dry leaves on large scale. Unsustainable practices such as complete removal of leaves, fruits etc. were adopted by the collectors during this period. These malpractices adversely impacting density and regeneration of the species in the natural forests of study area. However, natural forests are also prone to natural hazards *i.e.* fire, drought, heavy rains etc. Grazing and human interference and other biotic pressure are also responsible for its complete absence in the forests of Bhopal.

Discussion

Cissus quadrangularis and *Tylophora indica* occurs in special habitat conditions in the studied area. Presence of one or two plants in studied areas indicates suitability of site as well as geo-climatic conditions and moisture availability in the region that are suitable to the species. The plant density of *Cissus quadrangularis* is negligible in the natural forest areas of Bhopal district including Mandla (East), which are otherwise known for good medicinal plant bearing forests of Central India. Both plants are getting depleted in the natural forests because of excessive collection of medicinally usable parts (stem, leaves etc.), uprooting of whole plant, by the people involved in the plant collection. The current harvesting practice by uprooting whole plant in the natural forests was found to be unsustainable. Forest floor has no chance of receiving mature and viable seeds due to immature plant collection. This is the main reason for poor regeneration. This appears to support poor quality as well as quantity of both the species in all the selected districts.

The present study reveals that population density and regeneration are slightly better in the forest of Mandla (East), as compared to natural forests of Bhopal. Habitat suitability and less anthropogenic factors are the causes of its presence in forest Mandla, while natural forests are more disturbed due to human activities in Bhopal district. Khuroo *et al.*, (2005) highlights the role of protected areas in the conservation of biodiversity and reported that *Gentiana kurroo* is fast heading towards local extinction in the Kashmir Himalaya due to anthropogenic pressures. Earlier, the species is distributed throughout the region, but now currently represented by only a single wild growing population, found in a protected area.

Major threats to natural habitats are expansion of agricultural activities in forest land, over grazing by domestic animals and frequent forest fires. Human activities and market domestic household demand, competition among villagers to collect more leads to overharvesting of both the species in the study area. Such activities may be adversely affecting the plant population and saplings in the natural forests of selected districts.

The present harvesting system is ecologically and socio-economically unsustainable. Apart from destructive harvesting and lack of value addition, general absence of local level institutions deprive the collectors fair and just wages for their works. There is no mechanism to discourage premature harvesting. The transport of these species as a non-nationalized produce does not require any transit pass and therefore forest department does not have reliable statistics on the actual extraction. Ved *et al.* (2003) on the assessment basis on IUCN categories, considered *Curcuma* (03 species), *R. serpentina* under the category of critically endangered. Oudhiya, (2006) reported that due to over exploitation; the availability of this herb is decreasing in Orissa, Chattisgarh and adjoining states. He also felt need to promote its commercial cultivation to reduce the pressure on natural forests. Mishra (2001) reported destructive harvesting, immature collection by locals and declining population, density of *C. caesia* and *R. serpentina* in the natural forests of central India. Dhar *et al.*, 2000; Kala 2000, 2005; Dhar, 2002, Uniyal *et al.*, 2006 also reported population status and ecological limitations of endemic population to localized niches are factor responsible for extinction. Singh, *et al.* (2008), studied population status of threatened plants in different landscapes and reported lowest density of *Aconitum heterophyllum* due to high pressure of anthropogenic activities and habitat specific distribution.

Conclusion

The present condition of (*Cissus quadrangularis* and *Tylophora indica*) plants in the natural forests of Bhopal Division is very precarious and needs immediate attention not only for conservation but also for propagation. Present study clearly shows that the position of both in the natural forest areas is very alarming in terms of less number of climber/plants per hectare, and poor quality. Unsustainable harvesting method adopted by the collectors is responsible for reduced density and regeneration of this species in the natural forests. The natural regeneration is also adversely affected due to immature harvesting of fruits/seeds. The competition for early collection among the locals living around the forests was more intense. The current harvesting pattern indicates necessity to immediately stop such unsustainable harvests of the plant by the local people and contractors. For the threatened medicinal plant species like *Tylophora indica*, cultivation is a conservation option because the constant drain of material from natural populations is much higher than the annual productivity. If the demand for these plants can be met from cultivated sources then the pressure on the wild populations will be relieved. In this case, the need for strict conservation of remaining populations, improved security of germplasm *ex situ* and investment in selection and improvement programs is extremely urgent. Because of various commercial and medicinal uses of this species, the condition in the natural forests is very critical. If the present condition is continues, sooner the species will completely vanish from the natural forests of both the districts.

Acknowledgement

The author is thankful to Dr. R B Lal, Director, IIFM for his support and valuable guidance. The author are thankful to the Divisional Forest Officer (DFO) Bhopal and Mandla (East) forest division for providing field facilities and valuable suggestions. I am also thankful to the forest department, field staff of both the district for providing assistance during the experimental work.

References

Dhar U (2002). Conservation implications of plant endemism in high altitude Himalaya. Current Science 82(2): 141-148.

Dhar U, Rawal RS, Upreti J (2000). Setting priorities for conservation of medicinal plants: A case study of the Indian Himalaya. Biological Cons. 95: 57-65.

Kala CP (2000). Status and conservation of rare and endangered medicinal plant in the Indian trans-Himalaya. Biological Conservation. 93: 371-379.

Kala CP (2005). Indigenous uses, population density, and conservation of threatened medicinal plants in protected areas of the Indian Himalayas. Conservation Biology. 19: 368-378.

Kanjilal PC (1933). A forest flora for Pilibhit, Oudh, Gorakhpur and Budelkhand. Published by Superintendent, Printing and Stationary, United Provinces, Allahabad, India.

Kotwal PC, Mishra M (2007). Recent harvesting practices and causes of population decline of critically endangered species Malkangni (*Celastrus paniculatus*) in the two districts of central India. Vaniki Sadesh. Vol. 31 (1): 4-13.

Mishra M (2005). Present ecological status and impact of harvesting of gum from kullu (*Sterculia urens*) trees: A case of three districts in Orissa. Vaniki Sandesh. Vol. 29 (2): 14-19.

Mishra M, Kotwal PC and RP Mishra (2004). Ecological status of rare and important medicinal plant Kali musli (*Curculigo orchiodes*) in the tropical forests of central India. Vaniki Sandesh. Vol. 28 (2 & 3): 16-23.

Mishra M (2001). Harvesting practices and management of two critically endangered medicinal plants in the natural forests of central India. Proceedings of the International seminar on Harvesting of Non wood Forest Products. Held at Menemen-Izmir, Turkey. pp. 335-341.

Mishra M (2009). Sustainable harvesting and management of medicinal plants in the natural forests of central India- Issues and challenges. Souvenir & Abstracts. pp: 50-65. In advancing frontiers in plant science: present status & challenges for future. Seminar held on 17-18 August, 2009 at CMD college, Bilaspur, Chattisgarh.

Mishra M, Kotwal PC (2009). Sustainable management and conservation of biodiversity in the natural forests of central India: a case of two medicinally important species. In *"Sustainable Management & Conservation of Biodiversity"* (*Eds.*) Pandey, Shivesh P Singh and Rashmi Singh. pp: 69-80. Publ. by Narendra Publishing House, New Delhi.

Mishra M, Teki S (2007). Present harvesting practices of Siali leaves (*Bauhinia vahlii*). and its impact on plant density and regeneration in the natural forest of three districts of Orissa state. Journal of Tropical Forestry. Vol. 23(2): 12-24.

Mishra Manish, Kotwal PC (2007). Harvesting decline and economics of Baichandi (*Dioscorea daemona*) in the natural forests of central India. Flora and Fauna. An International Research Journal. Vol. 13(2): 243-248.

Mishra M, Kotwal PC (2009)Traditional harvesting and processing methods of *Dioscorea daemona* (Baichandi) tubers in the forests of Madhya Pradesh, India. Journal of Tropical Medicinal Plants (Malaysia). Vol. 10 (1): 113-118.

Oudhiya P (2006). Traditional Medicinal Knowledge about common herbs used for the treatment of Hydrocele in Chhattisgarh, India. Botanical.com. website viewed on 25/6/2006. http: // www.botanical.com/site/column_poudhia/ 66_hydrocele.html.

Prasad Ram, Kotwal PC, Mishra M (2002). Harvesting practices of Safed musli (*Chlorophytum* spp.) and its ecological impact on the natural forests of central India. Journal of Tropical Forestry. Vol. 18 No. (1): 9-24.

Prasad R, Patnaik S (1998). Conservation Assessment and Management Planning (CAMP) of NTFPs in Madhya Pradesh- Concept and Procedure. Workshop proceedings on "CAMP for NTFPs in M.P." held at IIFM, Bhopal, pp: 7-9.

Prasad R, Kotwal PC, Mishra M, Mishra RP (2002). Standardizing methodology for sustainable harvest of some important NTFPs in Madhya Pradesh. Research project report submitted to Indian Institute of forest Management (IIFM), Bhopal, Madhya Pradesh. pp.1- 97.

Prasad R, Kotwal PC, Mishra M (2002). Impact of harvesting of *Emblica officinalis* (Aonla) on natural regeneration, health, vitality and ecosystem in central Indian forests. Journal of Sustainable Forestry. Vol. 14 (4): 1-12.

Singh KN, Gopichand AK, Brij L (2008). Species diversity and population status of threatened plants in different landscape elements of Rohtang Pass, Western Himalaya. Journal of Mountain Science. 5: 73-83.

Uniyal SK, Kumar A, Brij L, Singh RD (2006). Quantitative assessment and traditional uses of high value medicinal plants in Chota Bhangal and Himachal Pradesh. Western Himalaya. Current Science. 91(9): 1238-1242.

Ved DK, Kinhal GK, Ravikumar M, Karnat RV, Sankar, Indresha JH (2003). Threat Assessment & Management Prioritization (TAMP) for the medicinal; plants of Chattisgarh & Madhya Pradesh. Medicinal plant taxonomy & distribution workshop proceedings. Proceedings of workshop FRLHT-IIFM, published by FRLHT, Bangalore, India.

Chapter 13

Pattern Diversity Assessment and Conservation of Pachmarhi, Madhya Pradesh

☆ *Sudhir Pathak*

Pachmarhi town and sanatorium in Sohagpur tehsil of Hoshangabad district is an isolated plateau in the mahadeo hill of the Satpura range, 32 miles from Pipariya station on the Itarsi-Jabalpur section of the Central Railway. The Plateau of the Satpura hills on which the town stands at an elevation of just 3,500 ft. has an area of 23 square miles, the greater part of which is covered with forest.

The total phytodiversity of Pachmarhi hill includes 1173 species. The pteridophytic flora includes 94 species. The hepatic flora includes 57 species. Beside these, 37 species of epiphytic mosses and 46 species of terrestrial mosses have been recorded. The angiospermic flora include 935 species and 4 species of Gymnosperms have been reported in the area.

The community of the area is a natural junction of two most important timber species *i.e.* Teak *(Tectona grandis)* and Sal *(Shorea robusta)*. The vegetational pattern of the area is basically dependant on the climate, soil and topographical features.

Materials and Methods

Extensive field survey and collection of angiospermic plants were done in Pachmarhi forest duning the rainy, winter and summer season because the time of flowering in plants was different. The identification of plants was done by the help of different floras.

Observations

Vegetational Patterns

Five types of vegetation have been recorded at Pachmarhi. These 5 types of vegetational patterns are due to differences in topographical situation, soil and microclimatic set-up.

1. The sal forest, dominated by *Shorea robusta* at the top of the hills.
2. The mixed evergreen forest in the middle zones of the hills and around the plateau of Pachmarhi dominated by *Mangifera indica, Terminalia tomentosa, Terminalia bellirica, Syzygium cumini* and *Anogeissus latifolia* species.
3. The mixed dry deciduoes forest dominated by good quality of teak at lower region and associated with *Albizia lebbeck, Chloroxylon swietenia, Pterocarpus marsupium* etc.
4. The grassland and medows vegetation on the flat plateau dominated by perennial grasses like *Andropogon pumatus, Heteropogon contortus, Themeda triandra, Cymbopogon martini, Pseudosorghum helepense* etc. Here grassland is developed due to secondary succession.
5. Dry thorn forest- this type of vegetation is formed on dry exposed rock where soil, is very poor. The species found here are mostly *Euphorbia, Manilkra hexandra, Rhus parviflora, Tecoma stans, Lantana camara etc.*

Pachmarhi, perhaps has one of the richest floras of India, representing north and south India floras. The rare bamboo species *Bambusa polymorpha* is present in certain area. The deep gorges below Pachmarhi plateau grow rare tree ferns *Cyathea gigentea, C. latebrosa* and the rare Pteridophyte *Psilotum nudum.* Certain moist banks of streams are covered with liverworts, mosses and ferns *Adiantum phillipense, Gleichenia linearis, Osmunda regalis* and others. In moist ravines, tree fern like *Cyathea gigentia* and *Cyathea letebrosa* add grace to surroundings. Sal forest are associated with *Terminalia tomentosa, T. chebula, Lagerstroemia parviflora, Sterculia villosa, Gardinia turgida, Manilkara hexandra, Careya arborea, Garuga pinnata, Mallotus phillippinensis etc.*

The ground cover is occupied by herbs like *Sida, Triumfetta, Crotolaria, Desmodium indigofera, Alycicarpus, Tephricia, Ageratum, Vernonia, Trichodesma, Ipomoea, Justicia, Leucas, Gompherna, Achyranthus, Euphorbia, Achypaha, Commelina.*

The common climbers around the plateau are *Bauhinia vahlii, Cissampelos, Argyria, Ipomea, Dioscorea, Gloriosa, Gymmnema, Clematis, Smilex, Pergularia* and *Ventilago* spp.

There are also few parasite like *Dendrophthae fulcata, Scurrula, Viscum, Cassytha,* and *Cuscuta* spp. Many epiphytic and terrestrial orchids like, *Vanda, Habenaria, Geodorum, Goodyera, Peristylus* etc. are common in this area. Moist humid and shady places are dominated by luxuriant growth of ferns and bryophytes *Osmunda regalis, Lycopodium voluble, Angiopteris evecta, Microlepia speluncae, Pteridium acquillineum, Nephrolepisacuta, Leucostegia immerse, Athyrium falcatum, Nephrodium indicum, Lastrea criocarpa, Polybotrea appendiculata, Blechnum orientale, Drynaria quercifolia* etc. along with some grasses and bryophytes.

Loss of Plant Biodiversity and its Causes

The forest from base town to Pipariya, till one reaches to the plateau at the top of Pachmarhi hills were very rich and thick under growth was very luxuriant and varied a few years before. But, today the whole composition of the hill has been changed.

The causes of the decline of the flora of this region are:

- ☆ Development of the hill in the light of tourism and tourist attraction of this place has been increased now a day in manifolds.
- ☆ Construction of forest road and approach road at various picnic spots.
- ☆ A large number of botanical tours visited these places from all parts of the country to collect the flora at this place.

- ☆ Collection of biological materials by biological local suppliers.
- ☆ Collection of medicinal plants by local people and also by various ayurvedic medicinal practitioners.
- ☆ Deforestation by local and tribal people in search of fire and fuelwood.
- ☆ Damage caused by introduction of exotic species *Lantana* sp. and *Parthenium* sp.
- ☆ Cutting and grazing by local people.
- ☆ Pollution of soil, water and atmosphere.

There are many species which are endangered or rare due to above activities.

Few Rare and Endangered Plants of Pachmarhi

In many places exploration on large scale has resulted in complete loss of many important species from this hill. Now in the present condition many species are not easily available in these area. Some important plant species which are very rare and endangered now-a-days are given in Table 13.1.

Table 13.1

Name of Species	*Present Status*
Angiospermic plants	
Clematis triloba	Endangered
Thalictrum foliosum	Endangered
Viola patrinix	Endangered
Strobilanthes callosus	Rare
Melastoma malbathricum	Endangered
Salix tetrasperma	Rare
Dillania pentagyna	Rare
Vanda spp.	Endangered
Drocera lanata	Rare
Lespedeza elegans	Rare
Dendrobium densiflorum	Rare
Balanophore diocia	Endangered

Contd...

Table 13.1–Contd...

Name of Species	*Present Status*
Pteridophytic plants	
Microsorium membranaceum	Rare
Pleopeltis macrocarpa	Rare
Lygodium flexuosum	Rare(extinct)
Cyathea gigentia	Extinct
Angiopteris evecta	Extinct
Polybotrya appendiculata	Endangered
Osmunda regalis	Rare
Thelypteris ciliata	Rare
Psilotum nudum	Extinct
Isoetes panchananii	Endangered
Gleichenia linearis	Rare
Botrychium virgianiarum	Extinct

Conservation of Plant Biodiversity

Conservation of biodiversity means taking steps to protect genes, species, habitats and ecosystems. This may be done by following steps:

- Biosphere reserves, National Parks, Sanctuaries and other types of resources should be created for *in situ* conservation.
- Habitat conservation is one of the top step for the conserving biodiversity.
- Reclamation of area by site plantations.
- Preservation of essential ecological processes which are responsible for maintain floristic richness of hill.
- Natural restocking of rare and endangered species.
- Propagation and multiplication of rare and endangered species by vegetative and tissue culture method.
- Sustainable utilization.
- Public meeting, discussion and symposia about importance of plants and their conservation shall be organized for Public awareness.

Acknowledgement

I thank the Department of Biotechnology New Delhi for funding this study and forest department for work permission in the Pachmarhi area. I am thankful to CCF Bhopal. I thankful to Prof. V P Singh (Former Head of *i.e.* MPS) Vikram University Ujjain (MP).

References

IUCN (1994). International union for conservation of Nature and Natural Resources. Published by IUCN gland, Switzerland.

Mukherjee (1984). Flora of Pachmarhi and Bori reserves. BSI. Kolkata.

Pathak Sudhir (2001). Plant diversity and community patterns of tropical evergreen forest Pachmarhi Hills (MP) PhD thesis, Vikram University Ujjain, MP.

Pathak Sudhir (2001). Plant diversity and community patterns of tropical evergreen forest Pachmarhi hills (M.P.) Ph.D. summary. Biosphere Reserves Information System (BRIS) Vol. Ist (No. 1) Bhopal.

Project Document (1996). Pachmarhi Biosphere reserve. Environmental Planning and Co-ordination Organization. Bhopal.

Singh VP Kaul A (2002). Biodiversity and vegetation of Pachmarhi Hills. Scientific Publishers, Jodhpur, India.

Singh VP, Kaul A, Singh N, Pathak SK, Sisodia IS (1998). Rare, endangered and threatened plants of Pachmarhi hills M.P. An ecological and biotechnological approach for biodiversity conservation. First technical report, Ist year DBT No. DBT/ R&D/19-03-1994.

Chapter 14

Sustainable Management of Avenue Trees: A Case Study Along Five Highways of Imphal West District, Manipur, India

☆ *Khomdram Nermeshori Devi*

Transportation corridors are major infrastructure elements of today's urbanized life. They are not simply thoroughfares for motor vehicles, but must also serve as public spaces where they have profound relationship with the environment (Dixon and Wolf, 2007). The core of the idea is that the construction and planning of roads and highways must encompass all the environmental, conservational, economic, social and aesthetic aspects suited to the community. Beautification of roadsides with appropriate trees is one such area. However, being a man-made corridor, the roadside vegetation is usually highly influenced by human choices.

A roadside tree, according to Maryland Roadside Tree Care Regulation (USA), is a plant that has a woody stem or trunk that grows all, or in part, within the surfaced side of a public road (Anon. 2005). In general, plantations of avenue trees along either side of roads are both for utility and aesthetic values offered by them. Trees with flower of varying colours blooming during a specific period enrich the beauty of the surrounding. Many workers are of the opinion that aesthetically pleasing roadsides offer high psychological satisfaction, reduced mental fatigue, lowered physical tiredness, higher concentration and mental alertness while driving (Randhawa, 1983; Mathur, 1993; Cumming, *et al.*, 2001 and Dixon and Wolf, 2007). In spite of anthropogenic influences while selection and plantation of varieties of flowers of different colours and hues present lively landscape along the roadside when in bloom.

Maintenance and management of roadside trees in cities of many advanced countries assume highly scientific and strong decision making approach. Cities like New York, Maryland (USA), Victoria (Australia), Cairo (Egypt), Singapore, etc. have some of the best planted avenues. These cities have very well-defined maintenance and management policies coupled with stringent legal support by way of appropriate legislation.

Area of Study

Imphal West is the state capital of Manipur in the North-Eastern India, having a total geographical area of 519 sq. km. It is located between latitudes 24.30°N to 25.00°N and longitudes 93.45°E to 94.15°E. The mean altitude of the District is 790m above MSL (Figure 14.1). The district enjoys a salubrious sub-tropical climate with an average rainfall between 1085 mm and 1434 mm. Summers are characterized by hot and humid rainy season whereas cool and dry period prevails during winter. The average temperature of the district varies from 0°C to 36°C (Laiba, 1992).

Imphal West District is connected with other places through three national highways namely National Highway No. 39, National Highway No. 53 and National Highway No. 150 and two state highways namely, Imphal Kangchup Road and Imphal Mayang Road. These five highways have a total length of 95.55 km within the district boundary of Imphal west. They are the most viable mode of movement and vital links in the lives of people. These highways

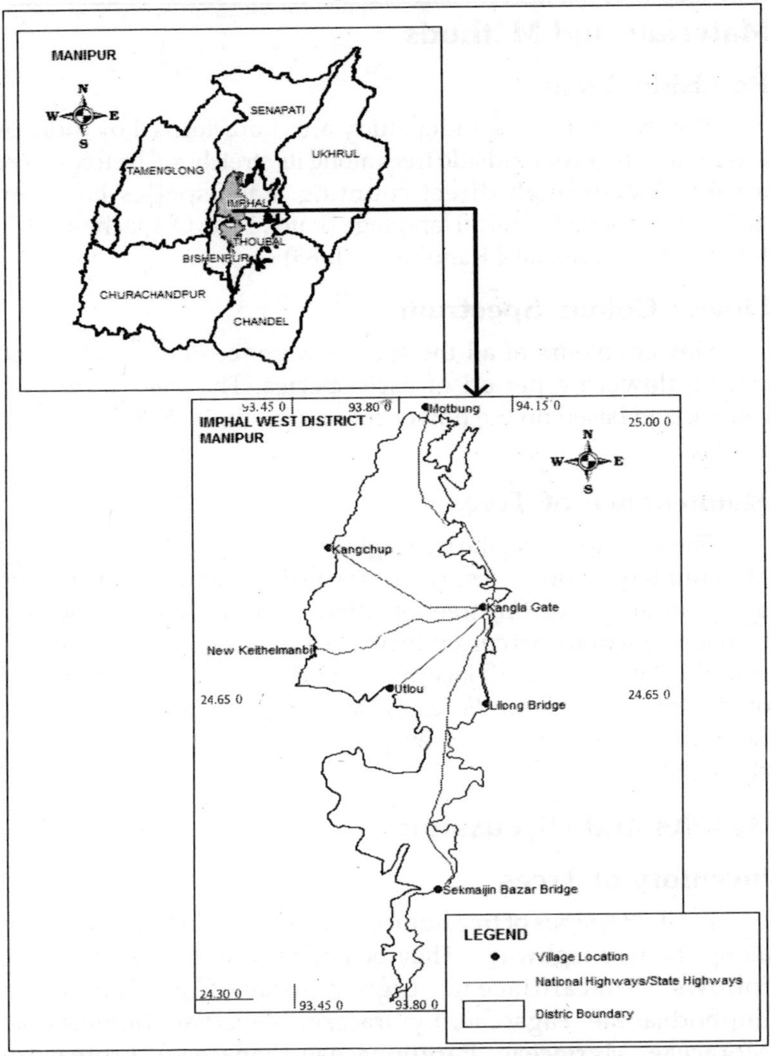

Figure 14.1: Location Map of Imphal West District in Manipur

are used under all weather conditions throughout the year. For the present study, observations were made between the years 2006 and 2008.

Materials and Methods

Roadside Trees

The five highways under study are characterized by more or less regular rows of roadside trees along its stretches. The trees were inventoried through direct counting. Each species has been described in semi technical language (Patel, 1968; Oomachan, 1977; Kanjilal *et al.*, 1982 and Randhawa, 1983).

Flower Colour Spectrum

Flower colours of all the species were recorded as observed during flowering period of each species. The trees were then categorized based on the flower colour (Singh, 1968 and Agarwal, 1973).

Maintenance of Trees

The avenues along the five highways in Imphal West District of Manipur have more or less regular roadside plantation comprised of mature and spreading trees of different nature. Maintenance and management practices currently practiced along the roads have been studied. Policies of two concerned departments of the Government of Manipur, namely, the Forestry Department and the Ecology and Environment Wing have been procured through proper official channels.

Results and Discussion

Inventory of Trees

In all, 23 species of tree numbering 4691 were observed to grow along the five highways. They belonged to fourteen families as follows: Anacardiaceae, Bignoniaceae, Caesalpiniaceae, Euphorbiaceae, Fagaceae, Lythraceae, Meliaceae, Mimosaceae, Moraceae, Myrtaceae, Papilionaceae (Fabaceae), Proteaceae, Rhamnaceae and Rubiaceae. Out of the total species, 16 were native and remaining seven species were exotic which are known to have high adaptability and tolerance. As regards the nature of the trees, twelve were deciduous, eight evergreen and three semi-evergreen (Table 14.1).

Among the species, number of *Acacia auriculiformis* and *Eucalyptus teretocornis* were found to be highest having 1296 and 899

Table 14.1: Classification of Trees Growing Along the Five Highways in Imphal West District, Manipur

Sl.No.	*Species*	*Family*	*English Name*	*Status*
1.	*Acacia auriculiformis* A.Cunn.	Mimosaceae	The Australian	Exotic
2.	*Albizzia lebbeck* (L) Benth.	Mimosaceae	East Indian Walnut	Native
3.	*Artocarpus heterophyllus* Lamk.	Moraceae	Jack-fruit tree	Native
4.	*Azadirachta indica* A. Juss.	Meliaceae	The Neem tree	Native
5.	*Callistemon lanceolatus* DC.	Myrtaceae	The Bottle Brush	Exotic
6.	*Cedrela toona* Roxb.	Meliaceae	The toon tree	Native
7.	*Delonix regia* (Boj) Raf.	Caesalpiniaceae	Flamboyant tree	Exotic
8.	*Emblica officinalis* Gaertn.	Euphorbiaceae	Indian Gooseberry	Native
9.	*Eucalyptus teretocornis* Sm.	Myrtaceae	Forest red gum	Exotic
10.	*Ficus benghalensis* Linn.	Moraceae	Banyan tree	Native
11.	*Ficus religiosa* Linn.	Moraceae	Peepal tree	Native
12.	*Grevillea robusta* Cunn.	Proteaceae	Silver Oak	Native
13.	*Jacaranda mimosifolia* D.Don.	Bignoniaceae	Jacaranda tree	Exotic
14.	*Lagerstroemiaspeciosa* (L.) Pers.	Lythraceae	The Queen's Flower	Exotic
15.	*Mangifera indica* Linn.	Anacardiaceae	Mango tree	Native

Contd...

Table 4.1–Contd...

Sl.No.	Species	Family	English Name	Status
16.	*Meyna laxiflora* Robyns.	Rubiaceae	–	Native
17.	*Parkia roxburghii* G. Don.	Mimosaceae	Tree Bean	Native
18.	*Pongamia pinnata* (L.) Pierre	Papilionaceae	–	Exotic
19.	*Pscidium guajava* Linn.	Myrtaceae	The Guava	Native
20.	*Quercus serrata* Murray	Fagaceae	Quercus tree	Native
21.	*Spondias mangifera* Willd.	Anacardiaceae	Plum mango	Native
22.	*Tamarindus indica* Linn.	Caesalpiniaceae	Tamarind	Native
23.	*Zizyphus mauritiana* Lamk.	Rhamnaceae	Indian Jujube	Native

trees respectively whereas *Emblica officinalis* and *Tamarindus indica* were least with 2 and 3 trees each respectively.

Flower Colour Spectrum

Based on the observations made, flower colours of all the 23 tree species recorded exhibit little variation. This may be due to the fact that the vegetation along the roadside is highly influenced by human choices. Along the highways, a total of 9 different flower colours were observed. The colours were namely blue, crimson, lilac, orange, pink, purple, scarlet, white and yellow. Among the trees, 'yellow' was the most abundant flower colour displayed by eight (8) different species. Next to it was 'white' found in six (6) trees; 'purple' in five (5) trees; 'orange' in four (4) trees; 'scarlet' in two (2) trees whereas 'blue', 'crimson', 'lilac' and 'pink' were observed in one species each. They were *Jacaranda mimosifolia, Callistemon lanceolatus, Pongamia pinnata* and *Lagerstroemia speciosa* respectively. During the observation, the flowering pattern showed that 'yellow' and 'purple' colours were longest dominant in nine different months starting from November. The period from March to May during pre-monsoon till the onset of monsoon shows maximum flower colour variation in which 'white' was the most predominant colour whereas it was 'yellow' during monsoon season. As regards winter season, 'purple' was the only colour visible during the season.

Maintenance and Management

Survival of roadside trees and thereafter their success lies in the single fact that they should be maintained and managed properly. There is a need of periodical vigilance of new saplings planted on a regular basis till they attain self-propagating stage. Among the mature trees, regular pruning and care is needed to prevent them from being died out or chopped off.

The long term success of tree planting along roadside depends on the quality of the ongoing maintenance and management of these trees. Roadside avenues in many advanced cities around the world present characteristics integral to the lifestyle and environment of the place. Cities such as New York, Maryland, Victoria, Chicago, etc. have well-defined ecologically sound and sustainable roadside landscapes. Their planning and design encompass aspects regarding the environmental, cultural, social and economic potentials that roadside trees can impart. Though relatively small, roadside trees have huge environmental benefits in moderating city environs, surface temperature maintenance, water conservation and ecosystem conservation (Nowak and McPherson, 1992; Wolf, 2004 and Dixon and Wolf, 2007). Besides, roadside trees can reduce highway maintenance costs as well as increase the aesthetics of the roadsides (Galvin, 1999 and Wolf, 2004). Moreover, these cities have stringent legal support for management right from the selection of species to replacement of dead and damaged trees (Anon., 2005 and Anon., 2006).

In the district of Imphal West, proper and systematic planting of roadside trees have been neglected and done in a haphazard and clumsy manner. Trees once grown are totally neglected after they attain maturity. The situation is almost the same for both the natural as well as planted trees along the roadside. The Forest Department at various occasions take up the plan of plantation and their maintenance but the survival rate has been found to be very low due to lack of proper management strategies. There has been complete lack of a well-defined scientific management plan for the highways. Not only, there are no legal frameworks to undertake the required strategies. Starting from the selection of species to the maintenance of mature trees, there hardly exists any systematic strategy along these highways. Thus, it is imperative to bring forth a proper and scientific management plan for roadside trees for a sustainable environment.

Conclusion

For any highway, well-maintained ecologically sound management plans for the roadside trees are necessary in order to derive the maximum environmental and aesthetic potentials of the trees. It is imperative to fully incorporate adequate financial sources as well as sound and scientific decision making while executing the strategies. There must also be a comprehensive legal framework so that it can equally bind everybody concerned to protect and nourish the roadside trees thereby ensuring the health and beauty of them towards achieving a sustainable environment.

Acknowledgement

I am very much thankful to late Prof. Sharda Khandelwal, Head, IGAEERE, Jiwaji University, Gwalior, late Prof. R. R. Das, Former Vice-Chancellor, Jiwaji University, Gwalior and Prof. W. Ingo Meitei,

Head, Horticulture Department, CAU, Imphal for the valuable technical and moral support rendered during the preparation of this material. Thanks are also due to the Director, IGAEERE, Jiwaji University, Gwalior for the laboratory and technical aid provided during the research. I am also grateful to Jiwaji University, Gwalior for the financial support in the form of research scholarship provided during the tenure of the work.

References

Anonymous (2005). Natural Resources Article: Roadside Tree Care. Department of Natural Resources, Maryland.

Anonymous (2006). Roadside Conservation Management Plans Guidelines. VIC Roads, City of Ballarat.

Cumming AB, Galvin MF, Rabaglia RJ, Cumming JR, Twardus DB (2001). Forest Health Monitoring Protocol Applied to Roadside Trees in Maryland. *Journal of Arboriculture* 27: 126-137.

Dixon KK, Wolf KL (2007). Benefits and Risks of Urban Roadside Landscape: Finding a Livable, Balanced Response. 3rd Urban Street Symposium. Seattle, Washington.

Galvin MF (1999). A methodology for assessing and managing biodiversity in street tree populations: A case study. *Journal of Arboriculture* 25: 124-128.

Kanjilal UN, Kanjilal PC, Das A (1982). *Flora of Assam* (Volume I, II and III). Government of Assam Publication.

Laiba MT (1992). The Geography of Manipur. Public Book Store, Imphal.

Mathur LM (1993). *Tree Plantation and Environment Awareness*. Ashish Publishing House, New Delhi-26.

Nowak DJ, McPherson EG (1992). Quantifying the Impacts of trees: The Chicago Urban Forest Climate Project.

Oommachan M (1977). Flora of Bhopal. JK Jain Brothers, Bhopal (India).

Patel R I (1968). Forest Flora of Melaghat. Bishen Singh Mahendra Pal Singh, Dehradun.

Randhawa MS (1983). Flowering trees. National Book Trust, India.

Singh JS (1968). Successional Variation in Composition, plant biomass and net community production in grassland at Varanasi. Ph.D. Thesis, Banaras Hindu University.

Statistical Booklet of Manipur Forest (2005). The Principal Chief Conservator of Forest, Government of Manipur.

Wolf KL (2004). Trees, Parking and Green Law: Strategies for Sustainability. USDA Forest Service, Southern Region Georgia Forest Commission.

Chapter 15

Certain Avenue and Roadside Trees in Morena City, Madhya Pradesh, India

☆ *R.P. Singh & Keshav Singh*

Trees are important to humanity not only economically and environmentally but also spiritually and aesthetically. They sustain human life through direct and indirect by providing a wide range of products for survival and prosperity.

Human are moving away from nature due to urbanization and industrialization. Avenue and roadside trees can help to make urban area 'green and beautiful'. Trees with colorful flowers or foliage add extra pleasant appearance.

Morena is a town in the state of Madhya Pradesh. It is the headquarter of Chambal division which is situated in the Chambal valley. Morena is situated between latitude 26°30′N and longitude 78°04′E at about 177 meter above mean sea level. The climate of

Morena city is one of the extreme types. The minimum temperature ranges between 2°C -5°C in December while maximum temperature ranges between 42°C-46°C during May/June.

Present work deals with the study of certain avenue and roadside plants in Morena city. Study is based on the short survey of the Morena city that is conducted during the year 2010.

Systematic Enumeration of Species

In the enumeration, the species are arranged alphabetically with name of the family, local name and flowering time.

1. *Acacia nilotica* (L.) Wild.

   Family- Mimosaceae; Local Name- Babul; Flowering- January

2. *Aegle marmelos* (L.) correa

   Family- Rutaceae; Local Name- Bel; Flowering- April

3. *Albizia lebbek* (L.) Benth.

   Family- Mimosaceae; Local Name- Siras; Flowering-April

4. *Azadirachta indica* A. Juss.

   Family- Meliaceae; Local Name- Neem; Flowering-April

5. *Bauhinia variegate* L.

   Family- Caesalpinaceae; Local Name- Kachnar; Flowering- February

6. *Delonix regia* (Boger ex Hook.) Rafin

   Family- Papilionaceae; Local Name- Gulmohar; Flowering- April

7. *Callistemon lanceolatus* DC.

   Family- Myrtaceae; Local Name- Bottle brush; Flowering- April

8. *Cassia fistula* L.

   Family- Caesalpinaceae; Local Name- Amaltas; Flowering- March

9. *Drypetes roxburghii* (Wallich) Hurusawa

   Family- Euphorbiaceae; Local Name- Putrajeeva; Flowering-March

10. *Eucalyptus maculata* Hook.
    Family- Myrtaceae; Local Name- Safeda; Flowering-Throughout the year
11. *Ficus religiosa* L.
    Family- Moraceae; Local Name- Peepal; Flowering-April
12. *Ficus benghalensis* L.
    Family- Moraceae; Local Name- Bargad; Flowering-Throughout the year
13. *Holoptelea integrifolia* (Roxbb.) Planch.
    Family- Ulmaceae; Local Name- Churel; Flowering-January
14. *Leucaena leucocephala* (Lamk.) de Wit.
    Family- Papilionaceae; Local Name- Subabul; Flowering-August
15. *Melia azadirach* L.
    Family- Meliaceae; Local Name- Bakain; Flowering-April
16. *Moringa oleifera* Lamk.
    Family- Papilionaceae; Local Name- Sahjan; Flowering-February
17. *Polyalthia longifolia* (Sonner.) Thw.
    Family- Annonaceae; Local Name- Ashok; Flowering-April
18. *Pongamia pinnata* (L.) Pierre
    Family- Papilionaceae; Local Name- Karanj; Flowering-July
19. *Prosopis juliflora* (Swartz) DC
    Family- Papilionaceae; Local Name- Vilayati babul; Flowering-February
20. *Syzygium cumini* (L.) Skeels
    Family- Myrtaceae; Local Name- Jamun; Flowering-February
21. *Tamarindis indica* L.
    Family- Papilionaceae; Local Name- Imli; Flowering-May

Result and Conclusion

21 species of tree were recorded in Morena city. They belonged to nineteen families as follows: Annonaceae, Rutaceae, Meliaceae, Mimosaceae, Papilionaceae, Caesalpiniaceae, Moraceae, Myrtaceae and Ulmaceae. Out of the total 21 species, 19 are native and remaining two species are exotic. In future, there is a need of extensive survey of this area to explore more species.

References

Bennet SSR (1987). Name changes in flowering plants of India and adjacent regions. Triseas Publishers. Dehra Dun, India.

Duthie JF (1960). Flora of Upper Gangetic Plains and of the adjacent Siwalik and Sub-Himalyan tracts. B S I, Kolkata. Vols. 1-3.

Maheshwari JK (1963). Flora of Delhi, CSIR, New Delhi.

Mudgal V, Khanna KK, Hajra PK (1997). Flora of Madhya Predesh. Vol. II, BSI, Kolkata.

Singh NP, Khanna KK, Mudgal V, Dixit RD (2001). Flora of Madhya Predesh. Vol. III, BSI, Kolkata.

Verma DM, Balakrishnan NP, Dixit RD (1993). Flora of Madhya Predesh. Vol. I, BSI, Kolkata.

Chapter 16

Study of Fish Biodiversity of Tighra Reservoir, Gwalior, Madhya Pradesh

☆ *R.K. Mahor, R.R. Kanhere & R.B. Gupta*

Water of any water body has its physico-chemical and biological characteristics. Water is more useful for animals and plants. Water covers about 71 per cent of earth surface. The biota of the surface water is governed by various environmental conditions. The water of Tighra reservoir is used for storage of fish, irrigation, industries, domestic and drinking purpose.

Present study was carried out for the study of fish diversity and physico-chemical parameters of water of Tighra reservoir Gwalior M.P. Similar attempts have been made in different fresh water body of M.P., India (Adoni 1985, Kataria *et al.*, 1996; Adolia 1991; Mahajan and Kanhere 1996; Sukand and Patil 2004; Ugale and Hiware 2005).

Study Area

The Tighra fresh water reservoir is situated about 20 Km West of Gwalior city at an altitude of 218.58m near SADA Magnet city.

Tighra Reservoir was constructed by Late Maharaja Madhav Rao Scindhia in 1910-1917 across a seasonal rainfall on Sank River near village Tighra, in Gwalior Tahsil.

The Tighra fresh water reservoir lies on 26-12′0″ latitude and 78-30-0″ E longitude. The Tighra reservoir is surrounded by hills from three sides. It is a large sized reservoir with water spread area of 2112 hectares. The average depth of the reservoir 24 meter. The reservoir is rain fed during mansoon period. The reservoir is also fed by Sank River.

Material and Method

The Standard method of APHA (1998) and Trivedi and Goel (1984) were used for study. For sampling of water reservoir was divided into 5 stations designated as $Z_1Z_2Z_3Z_4Z_5$ so that maximum area of reservoir was covered. The water sample was collected in Plastic cans having two litter capacities. The temperature and pH of water were measured at the research site by thermometer and Systronic Battery operated pH meter. The samples were stored in refrigerator until the test in laboratory.

Result and Discussion

Physico-chemical characters of Tighra reservoir are shown in Table 16.1.

The water temperature is an important factor which influences growth and spawning of organism. In present investigation the minimum water temperature was recorded during January (13°C) and maximum was recorded in the month of June (30.8°C).

Ayyappan and Gupta (1981); Vijay Kumar (1992) working on the limnology of the Ramasamunder tank and Jagath tank, Karnataka and many other workers have also observed similar trends in different water bodies (Singh and Swaroop, 1979). In present study, the pH has been recorded January to December 2008. During the study periods pH value was recorded 7.2 to 8.2. The pH values were alkaline throughout the study period. The values of transparency ranged between 130.20 cm to 224.40 cm. The minimum value of transparency was recorded 130.20 cm in the month of July while the maximum was recoreded 224.40 cm in the month of July.

Table 16.1: Physico-Chemical Parameter of Tighra Freshwater Reservoir Gwalior, Madhya Pradesh–2008

Parameter	*Jan*	*Feb*	*Mar*	*Apr*	*May*	*Jun*	*Jul*	*Aug*	*Sept*	*Oct*	*Nov*	*Dec*
Temperature °C	13	20.8	22.58	26.4	26.66	30.8	28.66	27.44	25.74	23.30	21.30	15.22
Conductivity mohms	370.4	402	312.8	212.2	347.2	469.2	960.8	287	397.8	749.6	386	374
pH	7.4	7.177	8.2	8.24	7.3	7.24	7.46	7.78	7.64	7.4	7.72	8.04
Turbidity mg/l	7.06	7.84	6.78	6.08	7.94	8.88	13.2	7.46	7.6	7.98	7.42	7.08
Carbonates bicarbonates and total alkalinity mg/l	142.6	204	187.6	213.08	120.8	143.6	113.26	118.2	135.6	140.8	144.2	143.8
TDS mg/l	349	316	315.8	252.4	243.4	316.8	613.4	513.4	483.4	468.2	412.2	392.8
DS mg/l	312.4	300.6	231.2	233	223.2	284.6	555.2	514.2	503.4	446.2	423.8	408.4
S.S. mg/l	45.8	44.4	42.8	38.4	23.8	27.8	42	44.2	37	25.6	30.8	29
Hardness Ca^{++}	56.8	28.64	27.94	33	25.6	30.66	33.51	25	23.26	34	56	57.4
Magnesium Hardness mg/l	68.52	83.84	52.8	53.2	72.6	66	65.64	68	71	68.4	83.2	86.8
Chloride mg/l	20.11	25.29	43.2	36.2	25.01	32.54	32.23	31.06	30.25	29.6	29.71	26.75
Nitrate mg/l	0.24	0.16	0.35	0.28	0.38	0.23	0.32	0.11	0.14	0.41	0.41	0.44
Sodium mg/l	8.08	8.0	8.2	8.06	7.1	9.16	8.46	8.24	8.34	8.34	8.0	8.08
Potassium mg/l	0.6	0.81	0.86	1.04	1.02	0.64	1.06	0.74	0.76	0.96	0.8	0.68

Contd...

Table 16.1–Contd...

Parameter	*Jan*	*Feb*	*Mar*	*Apr*	*May*	*Jun*	*Jul*	*Aug*	*Sept*	*Oct*	*Nov*	*Dec*
D.O. mg/l	8.48	7.52	7.12	6.42	6.22	5.98	7.62	6.62	6.62	7.98	7.62	7.78
BOD mg/l	1.78	2.6	3.36	3.22	3.72	3.74	2.44	2.5	3.2	3.46	2.66	2.76
COD mg/l	26.62	98	66.2	20.57	36.06	20.1	31.37	33.6	32.26	43.6	26.32	26.58
Transparency cm	162.20	186.60	211.40	220.20	224.40	179.60	130.20	133.20	133.80	147.20	148.80	182
Phosphate mg/l	1.21	1.24	1.25	1.27	1.35	1.26	1.06	0.68	0.80	0.70	1.0	1.56

The conductivity was recorded 212.2 mhos in the month of April and 960.8 mhos in the month of July.

Dissolved oxygen is one of the most important parameters of the water quality, affecting fauna & flora of the aquatic ecosystem. The minimum value was recorded 5.58 mg/l in the month of June and maximum 8.48 mg/l in the month of January. The dissolved oxygen concentration was recorded high during January 2008.

Table 16.2: Special Diversity of Fishes in Tighra Reservoir (January to December, 2008)

Sl.No.	*Month*	*Species Diversity*
1.	January	0.54
2.	February	1.37
3.	March	0.73
4.	April	0.81
5.	May	0.91
6.	June	0.51
7.	July	0.89
8.	August	1.1
9.	September	1.1
10.	October	1.15
11.	November	0.94
12	December	0.78

The alkalinity ranged between 113.26 to 213.8 mg/l. The minimum values of alkalinity was recorded 13.8 mg/l in the month of July and maximum 213.8 mg/l in the month of April 2008. The minimum values of TDS was recorded 243.4 in the month of May and maximum 613.4 mg/l in the month of July.

The minimum value of suspended solid recorded 23.8 mg/l in the month of May and maximum 45.8 mg/l in the month of January. The value of dissolved solids ranged between 243.4 to 61.3 mg/l. The minimum value of chlorides was recorded 20.11 mg/l in the month of January and maximum 43.2 mg/l in the month of March. The minimum value of nitrate was recorded 0.11 mg/l in August and maximum 0.44 mg/l in the month of December. The minimum value of BOD was recorded 1.78 mg/l in the month of

January and maximum 3.74 mg/l in May. The value of COD was recorded between 20.1 mg/l to 98 mg/l in the month of June and February respectively. The value of hardness of Na^+, Ca^{++}, Mg^{++}, K^{++} were found to be in permissible limits.

During the study period 32 Fish fauna were identified from the reservoir, which are catagorised into major carp (3), minor carp (13), cat fishes (13) and exotic fish (03). The maximum weight of dominant species of fish was observed as *Labio rohita*, 4 Kg. and minimum weight of *Labio calbasu* and *Catla catla* 1 kg each. Dominant species occurring throughout year were *Labeo rohita, Cirrhinus mirgala, Mystus seenghala, Catla catla, Labeo calbasu*. Species diversity of fishes were observed (1.37) to be higher in the month of February while lower was in the month of June 2008. The species diversity fluctuated monthly.

The study shows that physico-chemical properties of Tighra reservoir was observed in the permissible limit and 32 species of fishes were found on the reservoir.

References

Abdholia UN, Chakrabarthy A, Srivastava V, Vyas A (1990). Community studies on Manero bentheds with reference to Limnochemistry of Mana Sarovar Bhopal. J. Matcon. 3(2): 139-154.

Abdul AJ (2002). Evaluation of drinking water quality in Tiruchirapalli Ind. J. Envir. Health 44(2): 108-112.

APHA, AWWA and WPCF (1985). Standard method for the examination of water of wastewater, 16th edn., 1268.

Kataria HC, Quareshi HA, Iqbal SA, Shandilya AK (1996). Assessment of water quality of Kolar reservoir in Bhopal (M.P.) Poll. Res. 15 (2): 191-193.

Mahajan A, Kanhere RR (1995). Seasonal variation of a biotic factors of a fresh water pond at Barwani (MP) Poll, Res, 14 (3): 347-350.

Ramesh M, Saravanan M, Pradeepa (2007). Studies on the Physico-chemical characteristics of Singallunai lake, Coimbatore, south India. In proceeding National Seminar on Limnol. Maharana Pratap University of Agriculture Technology, Udaipur, India.

Shrivastava SK, Singh D, Prakash S, Ansari KK (2007). Studies on Physico-chemical parameters of distillery effluent and their correlation. Indian J. Applied & Pure Bi, 22(2): 231-234.

Sukand BM, Patil HS (2004). Water quality assessment of Fort lake of Belgaum (Karnataka) with special reference to Zooplankton, 25(1): 99-102.

Tewari DD, Mishra SM (2005). Limnological study during rainy season of Sectadvan Lake at Shrawasti District J. Ecophysiol. Occupant. Health 5: 71-72.

Trivedy RR, Goel PR (1986). Chemical and biological methods for water pollution studies environmental publication's Karnad, pp. 215.

Ugale BJ and Hiware CJ (2005). Limnological study of an ancient reservoir Jagtunga Samudra located at Kandhar, Dist. Nanded, Maharasthra State, India, Eco. Envi. And Cons. 11 (3-4):473-475.

WHO (1992). Guidelines for drinking water quality recommendations Word Health organization, Geneva.

Chapter 17

Conservation of Biodiversity Resources in Madhav National Park

Vishwambhar Prasad Sati

Biodiversity means number and variety of different organisms and ecosystems in a certain area. Organisms include plants, animals and micro-organisms. Resources can be defined as an inner ability or capacity that is drawn on in time of need, or such abilities considered collectively. Conservation on the other hand refers to sustainable use and protection of biodiversity resources. A campaign for the conservation of biodiversity resources mainly started after 1970 when the world community came to know about a large-scale loss of fauna and flora and it is continued until today but it has been too late to control over the situation.

Madhav National Park (MNP) was established in 1958. Earlier, it was known as Shivpuri National Park. Scindia rulers have used it as hunting spot. They also developed some places of interest within the park area. In a dense forest area, George Castle was constructed in 1911 to commemorate George V. Inside the Park Mahuwar River in 1918 is the major parts of MNP. Diversity in faunal and floral resources is tremendously high in the MNP. Large number of plants, birds, reptiles, and wild life species are found in this park. Currently, many of them are endangered and listed in a red book. The great Indian bustard (Son Chirraiya) is one of the endangered species. Conservation of all these biodiversity resources is indeed inevitable and a need of the hour.

Geographical Location

MNP is located about 3 Km away from Shivpuri town, surrounded by the villages. Previously its area was 165.32 sq km. At the latter stage, 181.3 sq. km area was included and now, the total area of the MNP is 346.6 sq. km. Table 17.1 shows the salient feature– latitude, longitude, altitude, average temperature, and average rainfall of the MNP. Average elevation of the MNP varies from 380-480 meter. Average temperature during summer is 39ºC and during winter is 16ºC, while average rainfall is 800 mm. Figure 17.1A shows geographical location of the MNP and Figure 17.1B shows spotted dear which are found tremendously in the park area.

Table 17.1: Salient Features of MNP

Latitude	25º23' to 25º33' N
Longitude	77º38' to 77º55' E
Altitude	380-480 M
Average Temperature	39º in Summer and 16º in Winter
Average Rainfall	800 mm

Source: Gathered from District Statistical Diary, Shivpuri 2007 and Toposheets.

Diversity Index

Diversity index of plant species was calculated. For calculating diversity index, a 1-ha plot (*i.e.* 100m × 100m) was established at the elevation of 420 m. The 1-ha plot was divided into 100 quadrates of 10m × 10m each. In each quadrate, data were gathered from trees and shrubs with diameter at breast height of the plants (dbh) of 1.0 cm and above. Other parameters recorded were species name and height of the plant. Data collected were classified and analyzed to calculate species diversity. About 3000 individual trees representing

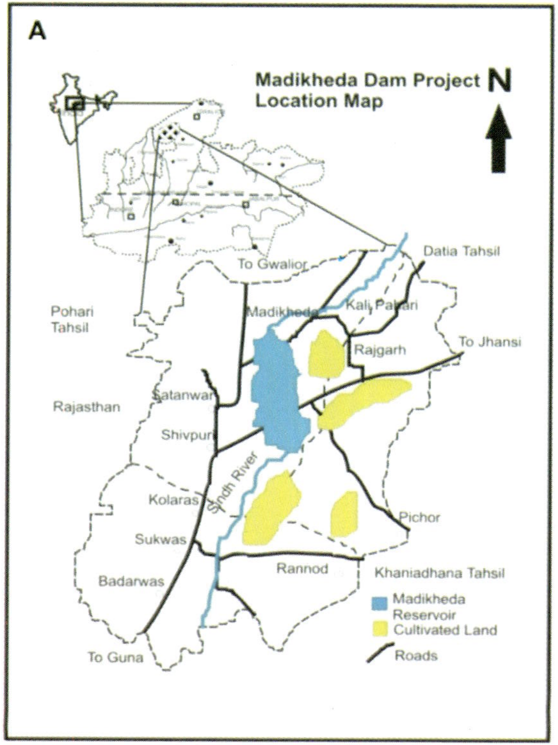

Figure 17.1A: Geographical Location of MNP

Figure 17.1B: Spotted Dear are Moving in the MNP

14 species were recorded in one hectare plot. Floral Diversity Index was calculated using Simpson's index of diversity:

$$D = \Sigma \pi_i^2$$

where

D: Simpson's index

π: Proportion of species

i: In the community

Diversity index of MNP was calculated = 0.876.

Forestland Use of the MNP

Table 17.2 shows forestland use in the MNP. Kardhai forest occupies 25 per cent of the total forest area followed by Khair 20 per cent. Grassland and Wasteland obtain equally 15 per cent forestland while others are 25 per cent.

Table 17.2: Forestland Use in MNP

Plant Species	*Area in per cent*
Kardhai	25 per cent
Khair	20 per cent
Grassland	15 per cent
Wasteland	15 per cent
Others	25 per cent

Source: Adapted from Sati (2001).

Depletion of Biological Resources in the MNP: Major Causes

There are various reasons for the depletion of biological resources in the MNP. These are:

1. Construction of Madikheda Dam
2. Stone mining in the Park and forested area
3. Illegal felling of trees and poaching of wildlife
4. Illegal grazing
5. Location of settlements and agricultural fields in the vicinity of park

Construction of Madikheda Dam

1. Due to dam construction heavy loss of forest was observed because, the reservoir submergence entailed loss of 3100.00 ha of forestland of which 1658.45 ha is a part of the Madhav National Park.
2. The effect of the construction of dam, by and large, on flora of the MNP and other areas is highly intensive.
3. The loss of fuel-wood due to submergence of MNP is 19686.7 m^3 costing Rs. 10.84 million.
4. A similar study for timber was carried out. The loss of Kardhai tree species was estimated as 581.22 m^3 with a value of Rs. 5.12 million at 1998 price level.
5. During construction stages, a large number of machinery and construction labour have mobilized. This activity has disturbed to the wildlife population. However, the magnitude of impact was difficult to predict.
6. Poaching was also increased during construction stage due to increased human interferences.
7. New roads were constructed or existing roads got upgraded. Workers cut trees for fuel wood and other requirements. This led to further degradation of forests.
8. The increase in pressure on existing forests is also likely to affect wildlife.

Stone Mining: Degradation of Forest and Decrease in Biodiversity

1. Mining and quarrying, either open cast or underground, destroys landscape and forest ecosystem.
2. The waste materials that remain after the extraction of usable ores are dumped on the surrounding land, thus causing loss of top soil, nutrients and supportive micro flora and vegetation.
3. As about 70 per cent stone mines are located in the park and forest area in Shivpuri District, forest depletion is tremendously high.
4. A large part of forestland in surrounding of mining areas is severely depleted.

5. Though, most of the mines in the forestland are closed but their impact can be noticed in the remnant areas.
6. Decrease in biodiversity – floral and faunal is also prominent as many species are endangered or in extinct.
7. In the MNP, Kardhai forest is abundantly spread and it is very useful species. But due to mining activity, about 30 per cent Kardhai forest is degraded.

Among the other causes, illegal felling of forest, poaching of wild animals, and location of settlements and agricultural land in the vicinity of the MNP are prominent. These causes are equally responsible for depletion of biodiversity resources in the MNP.

Conservation of Biodiversity Resources

There are various steps which can conserve biodiversity resources in the MNP. These are as follows:

1. Sustainable use and protection of biological resources
   (*a*) Rehabilitation of villages located in and surroundings of MNP in the other areas.
   (*b*) Similarly, agricultural land can also be shifted.
2. Reducing illegal cutting of trees and illegal grazing
   (*a*) A rational planning should be framed and implemented to stop cutting of trees and illegal grazing.
3. Protecting and preserving wild animals
   (*a*) Specific areas within the MNP should be identified where wild animals can be protected.
   (*b*) Wildlife act should be implemented properly to stop poaching of wild animals.
4. Stone mining located in and around MNP should be closed immediately.
5. The areas where stone mining has already operated, reclamation of degraded land should be ensured.
6. These majors will lead conservation of biodiversity in the MNP

Conclusion

The MNP is one of the biodiversity hotspot is in Madhya Pradesh. Meanwhile, the biodiversity resources are tremendously depleting. On one hand, the Forest and Environment Department and the Government of Madhya Pradesh are working together for conservation of the MNP, depletion of the resources in the MNP are very high due to the given factors, on the other. There is a need for conserving these resources that can sustain the livelihood of the rural people. One of the tools for conserving these resources is the local people's participation that must be ensured while framing and implementing any policy for the conservation of biodiversity resources.

Acknowledgement

I acknowledge to the UGC, Bahadur Shah Zafar Marg, New Delhi for funding me to commence this project.

References

Sati VP (2001). Environmental Impact Analysis and Socio-Economic Study of Madikheda Dam Project in Shivpuri district, unpublished report of MRP submitted to the University Grants Commission, New Delhi.

Index

M

N

O

P

R

S

T

U

W